Ion-Selective Electrode Methodology

Volume II

Editor

Arthur K. Covington

Reader in Physical Chemistry
University of Newcastle
Newcastle upon Tyne
England

CRC Press
Taylor & Francis Group
Boca Raton London New York

CRC Press is an imprint of the
Taylor & Francis Group, an informa business

First published 1979 by CRC Press
Taylor & Francis Group
6000 Broken Sound Parkway NW, Suite 300
Boca Raton, FL 33487-2742

Reissued 2018 by CRC Press

Library of Congress Cataloging in Publication Data

Main entry under title:

Ion-selective electrode methodology.

Includes index.
1. Electrodes, Ion selective. I. Covington, Arthur
Kenneth.
QD571.1578 543'.087 79-10384
ISBN 0-8493-5247-9 (v.1)
ISBN 0-8493-5248-7 (v.2)

A Library of Congress record exists under LC control number: 79010384

Publisher's Note
The publisher has gone to great lengths to ensure the quality of this reprint but points out that some imperfections in the original copies may be apparent.

Disclaimer
The publisher has made every effort to trace copyright holders and welcomes correspondence from those they have been unable to contact.

ISBN 13: 978-1-315-89477-5 (hbk)
ISBN 13: 978-1-351-07387-5 (ebk)

Visit the Taylor & Francis Web site at http://www.taylorandfrancis.com and the
CRC Press Web site at http://www.crcpress.com

PREFACE

The idea for a volume of ion-selective electrodes, hailed as an important advance in analytical chemistry, in CRC's UNISCIENCE Series arose out of the great interest shown in my article for CRC's *Critical Reviews in Analytical Chemistry* published in 1974 (*Crit. Rev. Anal. Chem.*, 1973, 3 (4), 355-406). In inviting experts to join me in the project, I was very much concerned to have those who were actively engaged in working with ion-selective electrodes and who could write on practical matters from firsthand experience. Their enthusiasm is apparent from the result which grew to two volumes.

The intention was to produce a book which perforce would never be far from the laboratory, although CRC's use of Handbook in another connection precludes our use of that word in the title.

I have, intentionally, wielded a strong editorial hand bringing, I hope, terminology and symbols to a common basis. I thank the contributors for their forebearance and their ready compliance with my suggestions. We trust you, the reader, will find these two volumes valuable in this second decade of ion-selective electrode potentiometry.

A.K.C.
Newcastle upon Tyne, England
January 1979

THE EDITOR

Arthur Kenneth Covington, B.Sc., Ph.D., D.Sc. (Reading) C.Chem. FRIC, is Reader in Physical Chemistry in the University of Newcastle upon Tyne, U.K. and is well known for his work on glass, ion-selective and reference electrode systems, as well as for his studies of electrolyte solutions by thermodynamic and spectroscopic methods. Dr. Covington represents The Chemical Society (London) on the British Standards Institution Committees on the pH Scale, pH meters and Glass Electrodes, is Principal U.K. Expert on, and Leader of, the International Standards Organization Working Group on "pH and Potentiometry", and Titular Member of IUPAC Commission V5 on Electroanalytical Chemistry. He co-edited *Physical Chemistry of Organic Solvent Systems,* Plenum, London, 1973 and *Hydrogen-Bonded Solvent Systems,* Taylor and Francis, London, 1968.

CONTRIBUTORS

D. M. Band, Ph.D.
Senior Lecturer
Sherington School of Physiology
St. Thomas Hospital Medical School
London, England

Richard P. Buck, Ph.D.
Professor of Chemistry
University of North Carolina
Chapel Hill, North Carolina

Peter Burton
Formerly Electronics Engineer
Electronic Instruments, Ltd.
Chertsey, Surrey, England

Arthur K. Covington, D. Sc.
Reader in Physical Chemistry
University of Newcastle
Newcastle upon Tyne, England

Robert W. Cattrall, Ph.D.
Senior Lecturer in Inorganic and Analytical Chemistry
La Trobe University
Bundoora, Victoria, Australia

Phillip Davison, Ph.D.
Assistant Analytical Superintendent
BP Chemicals, Ltd.
Salt End, Hull, England

G. J. Moody, Ph.D.
Senior Lecturer in Chemistry
University of Wales Institute of Science and Technology
Cardiff, Wales

Géza Nagy, Ph.D., C. Sc.
Senior Research Fellow
Institute for General and Analytical Chemistry
Technical University
Budapest, Hungary

E. Pungor, Ph.D., C. Sc.
Professor, Head,
Institute for General and Analytical Chemistry
Technical University
Budapest, Hungary

Malcom Riley, Ph.D.
Chief Chemist
Electronic Instruments, Ltd.
Chertsey, Surrey, England

R. J. Simpson
Senior Research Officer
SIRA Institute
Chislehurst, Kent, England

J. D. R. Thomas, D. Sc.
University Reader in Chemistry
University of Wales Instutute of Science and Technology
Cardiff, Wales

Klara Tóth, Ph.D., C. Sc.
Associate Professor
Institute for General and Analytical Chemistry
Technical University
Budapest, Hungary

T. Treasure, M.S., F.R.C.S.
Senior Registrar in Cardiac Thoracic Surgery
Brompton Hospital
London, England

Pankaj Vadgama, M.B., B.S., B.Sc., M.R.C. Path.
Medical Research Council Training Fellow
Department of Clinical Biochemistry
Royal Victoria Infirmary
Newcastle upon Tyne, England

TABLE OF CONTENTS

Volume I

Volume II

Chapter 1

GAS-SENSING PROBES

M. Riley

TABLE OF CONTENTS

I. INTRODUCTION

It is now more than 8 years since the appearance of the first commercial potentiometric gas sensor for ammonia. Similar sensors for other gases subsequently appeared, but the ammonia gas-sensing probe remains the most significant in terms of sales volume and now ranks alongside the best-selling ion-selective electrodes in this respect. These gas-sensing probes display outstanding selectivity by comparison with many ion-selective electrodes and are not restricted to the determination of gases; they are widely used for the determination of ions which can be converted to gases by appropriate chemical pretreatment, ammonium ion, for example, being determined as ammonia after making the solution alkaline and nitrite ion being determined as oxides of nitrogen after making the solution acidic.

In 1957, Stow et al.[1] first described the concept of determining the partial pressure of gas dissolved in a sample by measuring the pH of a thin film of solution separated from the sample by a hydrophobic, gas-permeable membrane. They constructed a probe to measure the partial pressure of carbon dioxide in blood, which was subsequently improved by Severinghaus and Bradley,[2] who considered the theoretical principles in more detail. About 10 years later, the previous work was reviewed by Severinghaus,[3] who reported the effects of various membrane materials and internal electrolytes, and by Smith and Hahn.[4]

Despite the widespread use of the so-called "Severinghaus" electrode in clinical medicine, little work was done to extend the principle to the determination of other gases. The situation might have been different if there had been a specific requirement for such determination in clinical practice, although it is now clear that the lack of the microporous synthetic polymers, suitable for use as membrane materials, which have become available only relatively recently, was at least a contributory factor in delaying progress. In any event, more than a decade elapsed before the next significant advance occurred, and probes for other acidic and basic gases began to make an appearance.

In 1973 Ross et al.[5] described several gas-sensing probes and their mode of operation and suggested several others which might be feasible. Bailey[6] has provided a comprehensive review of the field up to 1976; since then, no radical new developments have occurred, but the area of application of gas-sensing probes has been extended to the solution of a wider variety of analytical problems.

II. CLASSIFICATION

A. General

Strictly, these sensors are neither ion selective nor are they electrodes. They comprise an ion-selective electrode in combination with a suitable reference electrode to form a complete electrochemical cell, whose e.m.f. (electromotive force) is a function of the activity of the determinand gas in the sample. However, because it is the ionic activity of one of the components of the thin film of electrolyte which is directly measured, they are properly considered as part of the family of potentiometric ion-selective electrodes. Nevertheless, the name "gas-sensing probes" has been proposed[7] for these sensors in an attempt to resolve the nomenclature difficulty, and this name will be used here.

Two distinct types of gas-sensing probes are now in use; all the initial developments in the field were concerned with gas-sensing membrane probes, while the somewhat different gas-sensing probes without membranes appeared later.

The construction of a typical gas-sensing membrane probe is illustrated in Figure 1. This shows a probe based on a glass electrode with a slightly convex pH-sensitive tip,

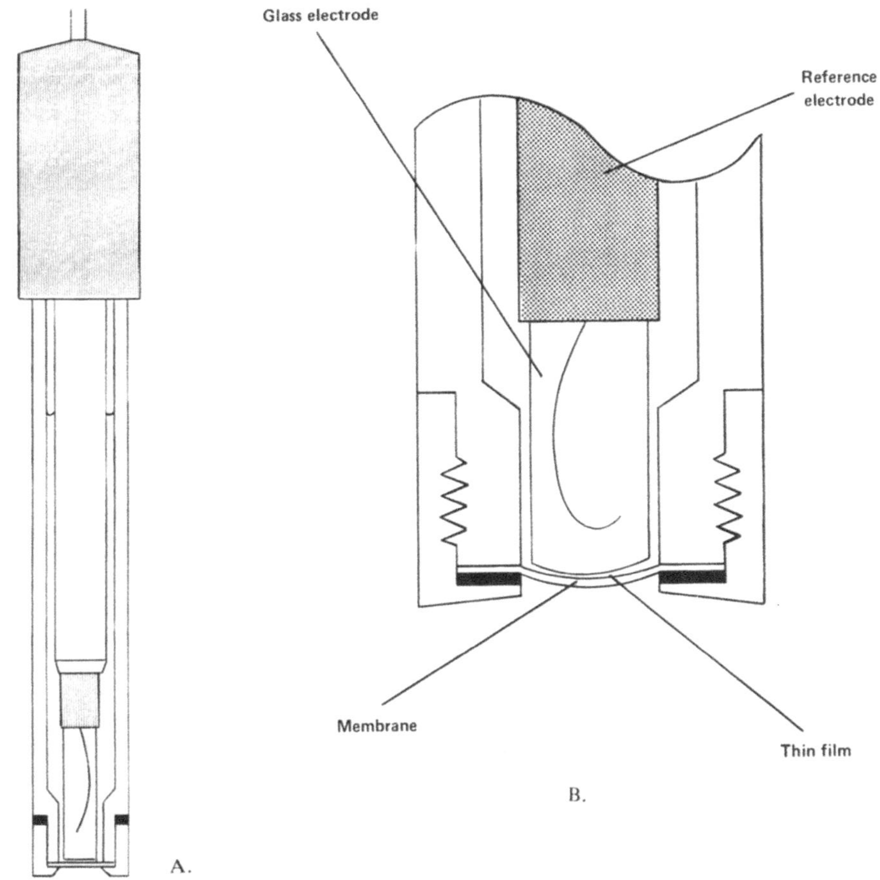

FIGURE 1. Construction of a typical gas-sensing membrane probe; A — overall layout and B — enlarged cross-section of sensing tip.

which is held against the gas-permeable membrane so as to sandwich a thin film of the internal electrolyte between itself and the membrane. On immersion of the probe in a sample, determinand gas diffuses through the membrane until the partial pressure of the gas in the thin film of electrolyte is equal to that in the sample. This equilibrium partial pressure of determinand gas determines the pH in the thin film, which is measured by means of the glass electrode and a suitable reference eletrode (commonly a silver/silver halide electrode) immersed in the bulk of the internal electrolyte.

Gas-sensing probes without membranes, otherwise known as "air-gap electrodes",[8] do not differ significantly in principle from gas-sensing membrane probes. An air gap several millimeters in thickness replaces the membrane, and the probe is suspended above the surface of the sample, which is contained in a special sealed vessel; equilibration between the sample and the thin film occurs by diffusion of determinant gas through the air gap. The thin film of electrolyte is applied, normally with a special sponge, to the pH-sensing membrane of the glass electrode prior to measurement; the electrolyte incorporates a suitable wetting agent to stabilize the film.

The relative merits of gas-sensing membrane probes and gas-sensing probes without membranes have been discussed previously,[7] and several benefits conferred by a membrane were enumerated. The arguments may be summarized by stating that gas-sensing membrane probes are, in general, more useful because they are easier to use but that "air-gap electrodes" are preferable for the analysis of samples which wet or otherwise impair the performance of membranes because the probe does not come into direct contact with the sample.

Although only gas-sensing probes incorporating glass pH electrodes have so far been considered, in fact it is possible to use other ion-selective electrodes as the basis of practical sensors. For example, as shown in Table 1, gas-sensing membrane probes for hydrogen sulfide and hydrogen fluoride are based on sulfide and fluoride ion-selective electrodes, respectively. Probes for these two gases could be based on pH electrodes instead but, as Table 1 indicates, in each case the internal electrolyte used with the preferred ion-selective electrode will effectively prevent interference by other volatile acidic or basic species in the sample, thus significantly enhancing selectivity. Table 1 lists the basic functional details of the best-known gas-sensing membrane probes; in principle, however, probes suitable for the determination of many other gases are feasible, and Ross et al.[5] have suggested several possibilities.

Whatever type of ion-selective electrode is used, it is clearly necessary that its configuration should be such as to allow a satisfactory thin film of internal electrolyte to be formed on its ion-selective membrane; in practice, this means that the sensing tip of the electrode will probably be flat or slightly convex. Thus, only a thin film of electrolyte will remain when the sensing tip is pressed against the gas-permeable membrane, minimizing the volume of solution which must equilibrate with the sample and, hence, the response time.

B. Membrane Materials

The synthetic polymeric materials commonly used to form a gas-permeable, hydrophobic membrane can be divided into two distinct groups, microporous and homogeneous.

A microporous membrane can be regarded as an inert, porous, polymeric matrix which merely acts as a support for a film of air through which the determinant gas diffuses; this air film is retained in the pores of the membrane material, which itself plays no significant part in the diffusion process. Polytetrafluoroethylene or polypropylene, having a porosity (void volume) in the region of 70% and a pore size of less than 1 μm, are the most commonly used materials for microporous membranes. Mem-

TABLE 1

Internal Electrochemical Systems of Some Gas-Sensing Membrane Probes

Determinant gas	Ion-selective electrode	Electrolyte solution[a]	Basic thin film reaction
NH_3	H^+	0.01—0.1 mol/ℓ NH_4Cl	$NH_3 + H^+ \rightarrow NH_4^+$
SO_2	H^+	0.01—0.1 mol/ℓ $NaHSO_3$ or $K_2S_2O_5$	$SO_2 + H_2O \rightarrow HSO_3^- + H^+$
H_2S	S^{2-}	pH5 citrate buffer	$H_2S \rightarrow S^{2-} + 2H^+$
CO_2	H^+	0.005—0.01 mol/ℓ $NaHCO_3$	$CO_2 + H_2O \rightarrow HCO_3^- + H^+$
NO_x	H^+	0.02—0.1 mol/ℓ $NaNO_2$	$2NO_2 + H_2O \rightarrow NO_3^- + NO_2^- + 2H^+$
HF	F^-	1.0 mol/ℓ H^+	$HF \rightarrow F^- + H^+$

[a] Silver/silver halide reference electrodes are commonly used; hence, electrolyte solutions are normally more complex than those shown.

brane thickness is usually about 100 μm or less, and in some cases the material is supported on an inert mesh or net, normally formed from a suitable polymer; in such cases, the mesh or net will be in contact with the sample rather than the internal electrolyte. "Air-gap electrodes" can be regarded as gas-sensing probes in which a relatively thick air layer serves as a membrane and are, thus, theoretically indistinguishable from gas-sensing probes with microporous membranes.

Homogeneous membranes, in contrast, can be regarded as solid polymeric films through which the determinant gas travels by the process of dissolving in the membrane, diffusing through it and coming out of solution on the other side. Silicone rubber is now the most widely used material for homogeneous membranes, although polythene and polytetrafluoroethylene have been used, and some of the newer fluorinated olefinic copolymers may be useful. Rather thin membranes are the rule, typical thickness being in the range of 10 to 25 μm, in order to obtain sufficiently fast response.

Microporous membranes, being essentially air membranes, are naturally very much more permeable to gases than homogeneous membranes, although there are wide variations in the permeability of a given membrane to different gases. This is illustrated in Table 2, which is based on figures given by Ross et al.,[5] where the figures in the last column are appropriate to microporous membranes or to "air gaps". The parameter Dk is directly proportional to the conventional "permeability coefficient", D representing the diffusion coefficient of the gas in the membrane phase and k representing the partition coefficient of the gas between the solution and membrane phases. Although the figures can only be approximate, they nevertheless indicate that even for silicone rubber the permeability to a given gas is less than that of a microporous membrane by a factor generally in excess of 1000.

It has been claimed[5] that, in general, the value of 10^9 Dk should be at least 10^3 cm^2/sec in order to obtain an analytically useful response time from a probe having a 100-μm thick membrane. In designing a probe for a particular gas, it would seem logical to select the membrane material which is most permeable to that gas in order to maximize the speed of response.

TABLE 2

Values of 10^9 Dk/(cm²/sec) for Diffusion of Various Gases Through Various Media

Gas	Low-density polythene	Dimethyl silicone rubber	Air
O_2	2.0×10^2	1.6×10^5	5.8×10^9
CO_2	3.8×10^2	2.9×10^4	1.6×10^8
H_2S		3.4×10^4	7.7×10^7
SO_2		3.8×10^3	3.7×10^6
NO_2		1.0×10^3	2.2×10^6
HF			9.9×10^5
NH_3		98	5.3×10^5
H_2O		5.1	1.3×10^2
HCl			2.3

In practice, some compromise may be necessary because the permeability of the membrane to water vapor may also be relatively high; this problem is discussed in Section III.F. In some cases, of course, membranes with sufficiently high permeability will not exist, but the advent of microporous membranes has greatly decreased this possibility.

III. FUNCTIONAL DETAILS

A. Response Mechanism

When a sample is presented to a gas-sensing probe, an equilibrium is eventually established by diffusion of determinant gas through the membrane or "air gap", when the partial pressure of the gas in the sample is equal to its partial pressure in the thin film of electrolyte solution on the sensing tip of the ion-selective electrode. The determinant equilibrates with the various constituents of the thin film, and the activity of one of these constituents is measured with the ion-selective electrode and reference electrode; this activity is thus related to the partial pressure, and hence to the concentration, of the determinant in the sample.

Taking an ammonia probe as an example, the ion-selective electrode is a glass pH electrode, and it is the pH of the thin film which is directly measured; the probe e.m.f. can be related to the ammonia concentration in the sample by considering the electrochemical cell on which the probe is based.

	Bulk internal electrolyte		Thin film		Glass electrode internal electrolyte	
Ag | AgCl	NH_4Cl (0.1 *M*) AgCl (s)	‖	NH_3 (aq) NH_4Cl (0.1 *M*) AgCl (s)	| Glass |	Na_2HPO_4 (0.075 *M*) KH_2PO_4 (0.175 *M*) KCl (0.08 *M*) AgCl (s) Gelling agent	| AgCl | Ag

The e.m.f., E, of this cell can be expressed[9]

$$E = E^\circ_{glass} + \frac{2.3RT}{F} \log a_{H^+} - E^\circ_{Ag/AgCl} + \frac{2.3RT}{F} \log a_{Cl^-} \quad (1)$$

where a_{H^+} is the hydrogen ion activity in the thin film and a_{Cl^-} is the chloride ion activity and is a constant. Thus,

$$E = E' + \frac{2.3RT}{F} \log a_{H^+} \quad (2)$$

The stability constant, K, for the formation of ammonium ions is

$$K = \frac{a_{NH_4^+}}{a_{H^+} \, p_{NH_3}} \quad (3)$$

and is sufficiently large that changes in the ammonium ion activity in the thin film, $a_{NH_4^+}$, due to changes in the ammonia partial pressure, p_{NH_3}, in the film can be neglected.[9] Thus, $a_{NH_4^+}$ can be considered to be constant and

$$E = E'' - \frac{2.3RT}{F} \log p_{NH_3} \quad (4)$$

From Henry's Law

$$p_{NH_3} = H \, [NH_3] \quad (5)$$

where H is a constant at a given temperature and a given total concentration of dissolved species.

At equilibrium, p_{NH_3} is the same in both the thin film and the sample; hence, the dependence of the probe e.m.f. on the ammonia concentration in the sample can be expressed by the Nernstian equation

$$E = E''' - \frac{2.3RT}{F} \log [NH_3] \quad (6)$$

Analogous equations can be derived for other gas-sensing probes, but for acidic gases, a positive sign replaces the negative sign in Equation 6.

B. Response Range

The upper and lower limits of Nernstian response are determined theoretically by the limits of the assumption, made in deriving Equation 6, that the activity of the ionic form of the determinant ($a_{NH_4^+}$ in the above derivation) is constant. For example, in the case of an ammonia probe, the upper limit is reached when the partial pressure of ammonia in the thin film is sufficiently high to produce, by hydrolysis, a significant increase in $a_{NH_4^+}$ in the film.

Hansen and Larsen[10] proposed a graphical method for predicting the response ranges of gas-sensing probes but implied that the theoretical lower limit is set by the concentration of determinant gas in the bulk internal electrolyte by claiming that, in the case of an ammonia probe, the pH in the thin film could not be lower than the pH in the bulk electrolyte. Bailey and Riley,[11] on the other hand, pointed out that there was a considerable amount of experimental evidence to the contrary and claimed that the thin film could be regarded as isolated from the bulk electrolyte; again using an ammonia probe as an example, they derived a theoretical calibration graph showing a much lower theoretical limit than that proposed by Hansen and Larsen. The essential features of this calibration graph are shown in Figure 2.

In practice, of course, the thin film cannot be completely isolated from the bulk internal electrolyte (otherwise, no current could flow between the two electrodes), and

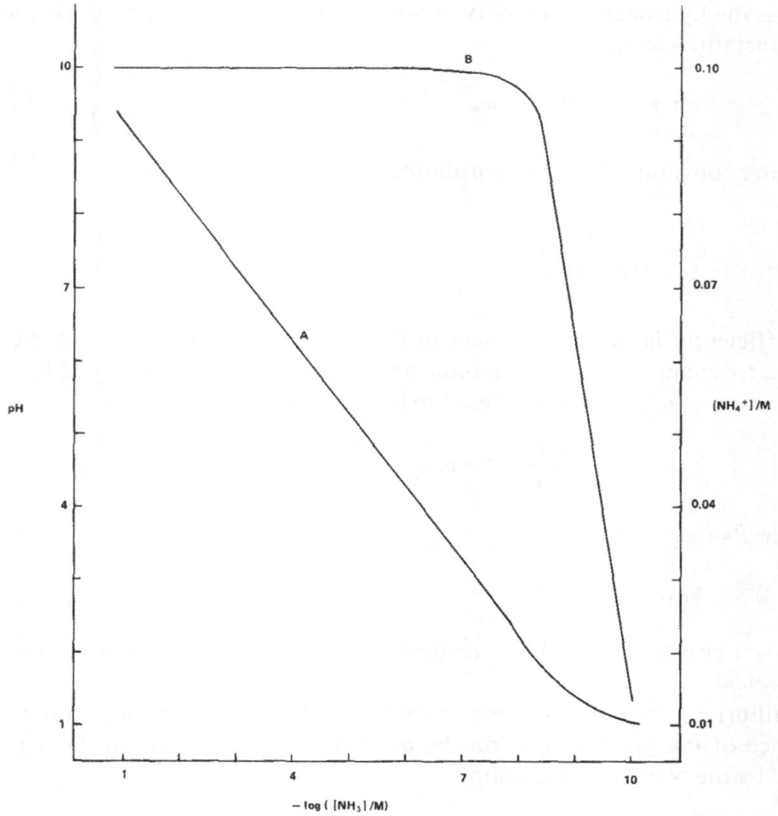

FIGURE 2. Theoretical graphs showing variation of pH and $[NH_4^+]$ in the thin film with sample $[NH_3]$ for an ammonia-sensing membrane probe with 0.1 mol/ℓ NH_4Cl internal electrolyte; A - pH and B - $[NH_4^+]$.[11]

some interdiffusion of the two solutions is inevitable. It follows that the limit of detection is represented by the concentration of ammonia in the thin film when equilibrium has been achieved between ammonia diffusing into the thin film from the bulk electrolyte and ammonia diffusing out of the thin film into a sample containing no ammonia. It also follows that the more restricted the interchange between the thin film and the bulk electrolyte, the closer will this limit of detection be to the theoretical limit and, conversely, the freer the interchange between the two solutions, the more the limit of detection will tend to be set by the concentration of ammonia in the bulk electrolyte.

The unknown rate of interchange between the thin film and bulk electrolyte solutions makes it difficult to predict limits of detection for gas-sensing probes, but it will seldom be the case that the measured detection limit is, in fact, that resulting from this interchange. The practical determination of limits of detection is often hampered by the experimental difficulty of producing water containing none of the determinant with which to prepare standard solutions and reagents. It is, for example, very difficult to obtain water containing less than 0.01 mg/ℓ of ammonia[9,12,13] or to completely exclude atmospheric carbon dioxide from samples, reagents, and apparatus.[14] An additional factor to be considered is the limitation imposed by the longer response times encountered at determinand concentrations close to the limit of detection; an extreme case is quoted by Gilbert and Clay[15] who found a response time of 145 min for an ammonia probe at a concentration of 6×10^{-8} mol/ℓ.

TABLE 3

Response Ranges of Gas-Sensing Probes

Probe	Type[a]	Nernstian response limit (mol/ℓ)	Limit of detection (mol/ℓ)	Upper limit (mol/ℓ)	Ref.
NH_3	M	ca. 10^{-6}	ca. 10^{-7}	1	5,7,9,12,13,15
	A	10^{-4}	10^{-5}		8
SO_2	H	5×10^{-5}	5×10^{-6}	5×10^{-2}	7
	M	5×10^{-6}	5×10^{-7}	10^{-2}	16
CO_2	H	10^{-5}	2×10^{-6}		14
	A	2×10^{-3}	10^{-4}	1	45
NO_x	M	2×10^{-6}	10^{-7}	10^{-2}	5,7
H_2S	M		10^{-8}	10^{-2}	5

[a] M = microporous membrane; H = homogeneous membrane; A = "air gap".

Table 3 lists the actual response ranges of some gas-sensing probes found in practice. The response range will depend to some extent on the concentration of the internal electrolyte solution,[3,5,10] a more dilute solution reducing the lower limits somewhat but also reducing the upper limit; in the case of the ammonia probe, the theoretical upper limit was reduced from 5 mol/ℓ for a 0.1 *M* ammonium chloride solution to 5×10^{-4} mol/ℓ for a 0.01 *M* solution.[10] The limit of detection obtained by Midgley[14] for a carbon dioxide-sensing membrane probe involved degassing of the acidic reagent and use of a closed flow system to exclude atmospheric carbon dioxide. Carbon dioxide is an interferent for the nitrogen oxide probe, and it seems likely that lower limits than those quoted could be achieved by rigorously excluding it.

The quoted upper limits are, in general, somewhat lower than the theoretically predicted values, and here again the interchange between the thin film and bulk electrolyte solutions is a factor. The longer a probe is exposed to a high determinant concentration, the more of the determinant will diffuse into the bulk electrolyte and the longer will be the subsequent recovery time before the probe gives a correct response in a sample with lower determinant concentration; in practice, increasing concentration hysteresis is observed as the determinant concentration approaches the concentration of internal electrolyte.

C. Selectivity

Compared to many ion-selective electrodes, gas-sensing probes show outstanding selectivity. They can only suffer direct interference from dissolved species in the sample which can both diffuse relatively rapidly into the thin film and either change the activity of the species sensed by the ion-selective electrode or interfere with the response of that electrode. Thus, ionic species in the sample cannot interfere, and this has been amply demonstrated in practice.

In the usual case, when the ion-selective electrode is a glass pH electrode, the only species which can interfere are those volatile species which can diffuse relatively quickly from the sample into the thin film and which also have an acidity or basicity comparable to or greater than that of the determinant. If permeability coefficient values are available, an estimate of the extent of interference can be made by comparison of the values of pK_a, permeability coefficient, and concentration for the interferent and the determinant. This will serve as a rough guide to the seriousness of the problem; it is, in any case, very difficult to predict quantitatively the extent of interference in particular practical situations because, for example, of the unknown rate of interchange between the thin film and bulk electrolyte solutions.

TABLE 4

Interference Data for Gas-Sensing Membrane Probes

Probe (type[a])	Interferent	Interferent concentration (mg/ℓ)	Determinant concentration (mg/ℓ)		Ref.
			Actual	Apparent	
NH_3(M)	Hydrazine	4	1	1.06	9
	Cyclohexylamine	4	1	1.08	9
	Morpholine	10	1	1.03	9
	Octadecylamine	0.4	1	1.14	9
	Methylamine	2.1	1.2	1.86	17
	Ethylamine	3.2	1.2	1.80	17
	Methanolamine	3.4	0.5	0.65	12
SO_2(H)	Acetic acid	6000	64	70.4	7
SO_2(M)	Acetic acid	300	64	70.4	16
	Hydrofluoric acid	60	64	70.4	16
NO_x(M)	Carbon dioxide	44	0.46	0.55	7
	Carbon dioxide	1320	46	50.6	16
CO_2(H)	Sulfite	100	10	14	14
		100	1	1.75	14

[a] M = microporous membrane; H = homogeneous membrane.

Table 4 gives interference data for several gas-sensing membrane probes and indicates that there are few interferents; these are generally volatile acidic or basic species, and their effects are essentially as would be predicted theoretically.

Drift of the probe e.m.f. is commonly found in the presence of interferents, and in cases of severe interference, very sluggish response may be encountered; in such cases, normal performance is most rapidly restored by establishing a fresh thin film of internal electrolyte solution. It follows that selectivity coefficients are not relevant to gas-sensing probes, and it is always better, if possible, to eliminate interferents by appropriate sample pretreatment than to attempt to correct the probe e.m.f. for their effects.

Substances which wet or coat the membranes of gas-sensing membrane probes sometimes cause problems in practice. Wetting agents, for example, will affect the hydrophobic properties of the membrane and ultimately the hydrophobicity will be destroyed and the membrane will become permeable to water; careful washing and drying may be effective in restoring hydrophobicity. Substances which coat membranes with films, gels, precipitates, etc. may reduce the permeability and eventually block the membrane surface completely. If such a problem cannot be solved by appropriate sample treatment, periodic cleaning to remove the coating will often allow satisfactory performance to be obtained. "Air-gap electrodes", which are suspended above the sample out of direct contact, are unaffected by membrane wetting and coating problems.

D. Response Time

Like ion-selective electrodes, gas-sensing probes exhibit faster response at high determinant concentrations, becoming more and more sluggish as the limit of detection is approached. Ross et al.[5] proposed a steady-state model which is useful in describing the response time characteristics of gas-sensing membrane probes in terms of membrane parameters, internal electrolyte composition, thickness of the thin film, and experimental conditions. From the model, they were able to derive an equation expressing the response time in these terms, and making certain assumptions, the equation was solved to yield the simpler equation

$$t = \frac{dm}{Dk}\left(1 + \frac{dC_B}{dC}\right) \ln \frac{(C_2 - C_1)}{EC_2} \qquad (7)$$

where t = time taken to achieve a given fraction (1 − E) of the total e.m.f. change due to a change in determinant concentration from C_1 to C_2; d = thickness of thin film; m = membrane thickness; D = diffusion coefficient of determinant in membrane; k = partition coefficient of determinant between sample or thin film and membrane phases; C_B = total concentration of all forms, other than gaseous, of determinant in internal electrolyte; C = concentration of gaseous form of determinant in internal electrolyte; and E = fractional approach to equilibrium = $(C_2 - C)/C_2$.

Equation 7 predicts that for increases in determinant concentration from low to high ($C_2 >> C_1$), the response time is essentially independent of the magnitudes of the concentration change, of C_1 or of C_2. It also indicates that response times for concentration increases are much shorter than those for concentration decreases; for example, the response time (E = 0.01) for a concentration change from 10^{-5} to 10^{-1} mol/ℓ determinant was calculated[5] to be about 13 times less than that for the reverse change.

The response times determined by Bailey and Riley[7] for three gas-sensing membrane probes are shown in Table 5; response time was defined as the time taken for the probe e.m.f. to reach a value 1 mV from its final equilibrium value following the concentration change. These results show that, for ammonia, the response times for decadic concentration increases were independent of the final concentration when this exceeded 10^{-4} mol/ℓ, as predicted by Equation 7. The other prediction of Equation 7 is also supported, the results clearly showing that the response times for increases in concentration were considerably shorter than those for decreases in concentration.

Gilbert and Clay[15] found that the response time of an ammonia probe increased from less than 1 min at 10^{-4} mol/ℓ to 7 min at 10^{-6} mol/ℓ and 145 min at 6×10^{-8} mol/ℓ for successive concentration decreases obtained by buffering samples with respect to ammonia (i.e., by appropriate adjustment of the pH of ammonium chloride solutions). Response time results, obtained by more conventional methods and reported by other authors,[12,13] are in satisfactory agreement. The e.m.f. of an ammonia-sensing "air-gap electrode" was reported[8] to reach equilibrium in 1 to 2 min in the

TABLE 5

Response Times of Gas-Sensing Membrane Probes

Final determinant concentration (mol/ℓ)	Time (sec)[a]		
	NH_3	SO_2	NO_2
For Tenfold Increases in Concentration			
10^{-2}	31, 30	32, 27	
10^{-3}	31, 31	36, 36	34, 38
10^{-4}	32, 33	425, 335	95, 92
10^{-5}	105, 110		370, 480
For Tenfold Decreases in Concentration			
10^{-3}	32, 30	68, 42	
10^{-4}	66, 34	155, 155	125, 66
10^{-5}	120, 70		460, 410

[a] Results of duplicate determinations are given.

concentration range 3×10^{-5} to 6×10^{-4} mol/ℓ, while a carbon dioxide-sensing probe of the same type was reported[8] to have a 95% response time of 30 sec in the concentration range 3×10^{-3} to 5×10^{-2} mol/ℓ. The time for a carbon dioxide-sensing membrane probe used in a closed flow system to reach its equilibrium e.m.f. following a tenfold concentration increase was found[14] to be 3 to 7 min for carbon dioxide concentrations above 1 mg/ℓ; the times included about 2-min "wash-out" time of the flowcell, and those for concentration decreases were found to be approximately twice as large, although as long as 30 min at concentrations below 1 mg/ℓ.

E. Temperature Effects

Although it is, in general, an unrealistic situation, it is possible to determine overall temperature coefficients for gas-sensing probes by allowing the system (probe and sample) to equilibrate at a given temperature, then changing the temperature and waiting for equilibrium to be reestablished. Experimental results obtained in this way, by warming up probes and samples in temperature-controlled cabinets, are given in Table 6; the values shown represent the algebraic sum of the temperature coefficients of all the temperature-sensitive elements of the system.

Because gas-sensing probes represent relatively complex systems, the effects of temperature on them are correspondingly difficult to characterize. For example, the different parts of the electrochemical cell shown in Section III.A all have different temperature coefficients and different thermal capacities; hence, the overall temperature coefficient will apparently vary with time until complete thermal equilibrium is attained following a temperature change.

In practice, the predominant temperature effect is often the osmotic transfer of water resulting from a difference in temperature between the thin film and sample solutions. If the sample temperature changes, then the temperature of the thin film will follow more slowly, because of the insulating effect of the membrane and the relatively large thermal mass of the ion-selective electrode with which it is in contact. The partial pressures of water vapor will thus be different on the two sides of the membrane, and water vapor will diffuse through the membrane so as to equalize the partial pressures, causing the probe e.m.f. to drift. As the temperature of the thin film slowly approaches that of the sample, the rate of water vapor transfer decreases and then when the temperatures are equal reverses in direction so as to restore the original electrolyte concentration in the thin film, causing the probe e.m.f. to drift back in the opposite direction.

TABLE 6

Temperature Coefficients of Gas-Sensing Membrane Probes

Probe	Temperature range (°C)	Temperature coefficient (mV/°C)	Ref.
NH_3	16—29	1.5	9
NH_3	28—35	1.3	7
SO_2	26—43	0.5	7
NO_x	25—31	0.2	7
NO_x	31—45	0	7
CO_2	15—35	1.0	14

This type of behavior was found[7] for an ammonia-sensing probe and is illustrated in Figure 3, which also shows the behavior of a sulfur dioxide-sensing probe under similar conditions. The latter probe had a homogeneous membrane, in contrast to the microporous membrane of the ammonia probe, so the rate of water vapor transfer was much smaller, resulting in a smooth curve without an "osmotic" trough. Similar behavior to that shown for the ammonia probe has been observed by other authors.[14,15]

It is clearly important when using gas-sensing probes with microporous membranes, or "air-gap electrodes", to ensure that the temperatures of the probes and of the samples are both adequately stable and, preferably, closely similar. It is recommended, for example, that probes should not be used in direct sunlight.

F. Osmotic Effects

If the total concentrations of dissolved species in the solutions on the two sides of the membrane are different, then an osmotic pressure difference results and water vapor will diffuse through the membrane until the activity of water is the same on each side. Also, as indicated in the previous section, an osmotic pressure difference results if there is a temperature difference between the solutions on the two sides of the membrane; this can lead to very large rates of water vapor transfer compared to those resulting from differences in the total concentration of dissolved species.[5]

Transfer of water across the membrane results in dilution or concentration of the internal electrolyte in the thin film, which causes the probe e.m.f. to drift. The rate of drift depends on the rate of water transfer, which is determined by the osmotic pressure gradient and the permeability of the membrane to water vapor, and although equilibrium will, in principle, eventually be reached, this is seldom observed in practice. Times in excess of an hour may be required. When the rate of water transfer is very low, the interchange between the thin film and bulk electrolyte solutions may be sufficiently rapid to reduce the dilution or concentration effect to negligible proportions. An indi-

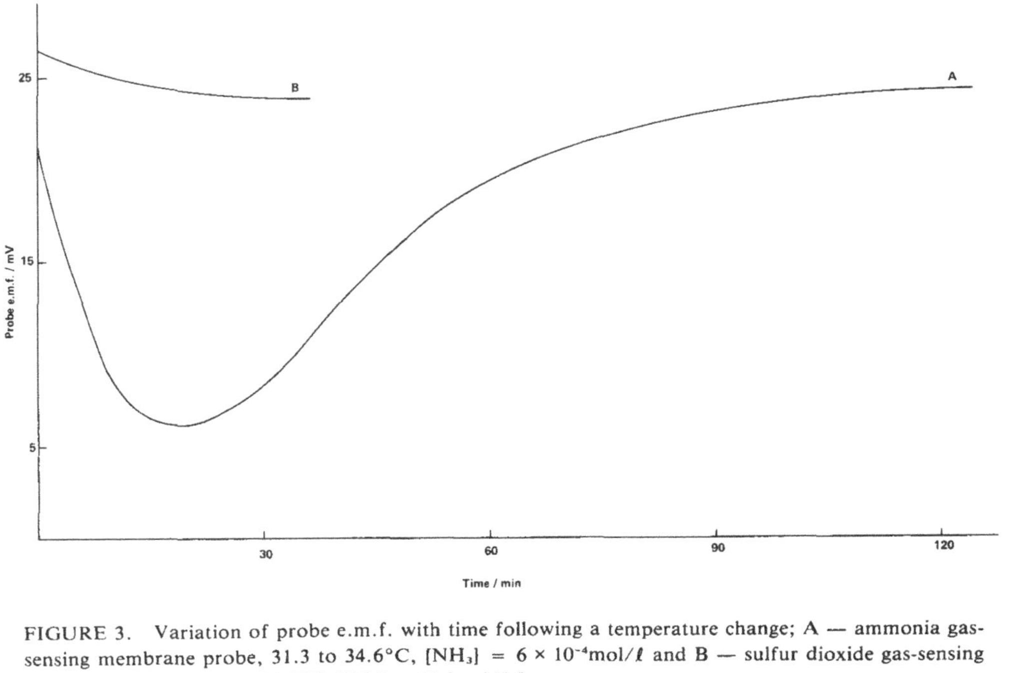

FIGURE 3. Variation of probe e.m.f. with time following a temperature change; A — ammonia gas-sensing membrane probe, 31.3 to 34.6°C, $[NH_3] = 6 \times 10^{-4}$mol/ℓ and B — sulfur dioxide gas-sensing membrane probe, 31.0 to 36.7°C, $[SO_2] = 10^{-3}$mol/ℓ.[7]

cation of the magnitude of osmotic effects is given in Table 7, which shows the initial rates of drift of probe e.m.f. found[7] for three gas-sensing membrane probes when used in determinant solutions containing sodium chloride at a concentration of 3 mol/ℓ.

The sulfur dioxide probe, with its homogeneous membrane, is again seen to be insensitive to osmotic effects. In the case of the ammonia probe, the drift rate was found[7] to decrease steadily over a period of 2 hr, and, subsequently, normal performance was restored immediately when a new thin film of internal electrolyte was created by loosening and then retightening the glass electrode.

In order to minimize osmotic effects, it is clearly desirable to use a membrane with as low a permeability to water vapor as possible; however, as indicated in Table 2, a decrease in the Dk value for water vapor will be accompanied by a decrease in the Dk value for the determinant gas, and thus the response time of the probe will be increased, possibly so much so that it is no longer a useful analytical device. Thus, a compromise is necessary in the selection of a suitable membrane material in order to minimize both osmotic effects and response time. In many cases, no real choice exists because only microporous membranes have sufficiently high permeability to the determinant gas. However, for example, the permeability of silicone rubber to sulfur dioxide is large enough to allow its use as the membrane of a sulfur dioxide probe, where it confers the benefit of insensitivity to osmotic effects.

Where osmotic effects cannot be eliminated by choosing a suitable membrane, there are two other alternative methods of preventing those due to difference in the total concentration of dissolved species between the sample and thin film solutions. The basic method is to match the osmotic strengths of the two solutions by diluting the sample, by selecting appropriate concentrations for the sample pretreatment reagents, or by adjusting the osmotic strength of the internal electrolyte. In certain cases, for example in the analysis of digests from Kjeldahl total nitrogen determinations with an ammonia probe, only the last of these will generally be practicable. These considerations apply equally to gas-sensing probes with microporous membranes and to "air-gap electrodes".

IV. PRACTICAL USAGE

A. Method Development

In the development of an analytical procedure based on a gas-sensing probe, a number of factors need to be considered. Eight important factors can be listed, and these

TABLE 7

Osmotic Drift of Gas-Sensing Membrane Probes in 3 mol/ℓ NaCl Solutions

Probe	Determinant concentration (mol/ℓ)	Initial drift rate[a]/mV/min
NH_3	5×10^{-4}	−12, −19, −14, −22
SO_2	1×10^{-3}	0.00, 0.00, 0.01, 0.00
NO_x	1×10^{-4}	−2, −2

[a] Results of replicate determinations are given.

are discussed below in two categories, sample pretreatment and standardization. All these factors must be considered, but, nevertheless, the resultant procedure will usually be simpler and more rapid than other, more conventional, procedures.

1. Sample Pretreatment

Sample pH—This should be adjusted, if necessary, so that essentially all of the determinant is present as dissolved gas; for ammonia and sulfur dioxide probes, for example, the sample pH should be above 12 and below 0.7, respectively. It may often be satisfactory, if maximum sensitivity is not required, to buffer samples to a pH where only a proportion of the determinant is present as dissolved gas, but this is generally less convenient and requires precise control of pH for accurate results.

Sample temperature—The temperatures of samples, including standard solutions, should be closely similar, ideally within 0.5°C, so that the Henry's Law constant remains unchanged, and the thin film of internal electrolyte in the probe remains at constant temperature.

Sample osmotic strength—This is particularly important for gas-sensing probes with microporous membranes and "air-gap electrodes". If necessary, the osmotic strengths of the sample and internal electrolyte solutions should be matched. In the usual case, when the sample osmotic strength is too high, it is best to dilute the sample if the determinant concentration allows this; otherwise, the osmotic strength of the internal electrolyte solution should be increased by addition of an inert electrolyte such as potassium sulfate or sodium chloride. For an ammonia probe with a 0.1 mol/ℓ ammonium chloride internal electrolyte solution, for example, matching of osmotic strengths will be required if the total concentration of dissolved species in the sample exceeds 0.3 mol/ℓ. In other cases, where the osmotic strength of the sample is too low, it will usually be possible to increase it by using an appropriate pretreatment reagent.

Complexation—In some samples, a significant proportion of the determinant exists in the form of complexes; ammonia, for example, may be present largely as complexes with metals such as copper and zinc. If it is required that this complexed determinant also be measured, then it must first be decomplexed; in the example quoted, the addition of EDTA to the pretreatment reagent serves to destroy the ammonia-metal complexes.

Interferents—If possible, any known interferents should be eliminated by appropriate pretreatment.

Determinant volatility—In general, it is best to make measurements in closed systems to prevent loss or gain of determinant to or from the atmosphere; the magnitude of the problem will depend on the volatility of the determinant, its concentration, and its abundance in the atmosphere in the environment in which measurements are made. Satisfactory ammonia measurements can often be made without using a closed system, but for laboratory measurements, it is a simple matter to fit a suitable O ring round the probe body, which serves to both seal the neck of the sample container and support the probe in the container.

Sample stirring—All samples should be stirred, taking the usual precautions to avoid temperature increases due to heat generated by the stirrer motor. In some cases, vigorous stirring is beneficial in minimizing the build-up of solid particles on the probe.

2. Standardization

The treatment of all samples, including standard solutions, should be identical, and, in particular, the total dilution of all samples should be the same. It is obvious that the accuracy of measurements can be limited by the accuracy of the standard solutions used; in some cases, such as sulfur dioxide measurement, it may be necessary to standardize the stock standard solution by an independent method immediately before use.

In the case of sulfur dioxide, the reagents available for preparation of standard solutions, such as sodium sulfite or potassium metabisulfite, are relatively impure (normally not better than 96% pure), and, in addition, the resulting solutions are relatively unstable.

B. Manual Analysis

In almost all respects, gas-sensing probes are used analytically in ways identical to ion-selective electrodes. Like ion-selective electrodes, they can be used for direct measurements, in titrations, and in the various known-increment methods. In general, they have the advantage of better selectivity but involve consideration of the additional factors of determinant volatility and osmotic strength.

C. Continuous Flow Analysis

1. Laboratory Analysis

Gas-sensing probes can conveniently be used in continuous flow analytical systems, which has the added advantage of minimizing any problems due to determinant volatility. Excellent results have been obtained[7] for the analysis of discrete samples with ammonia and sulfur dioxide-sensing membrane probes, at rates of 60 and 30 samples per hour, respectively, using the flow system depicted in Figure 4.

The probes were fitted with flow-through and caps,and debubbling of the liquid stream was not necessary—probably because the probes are complete electrochemical cells, and no external reference electrode, which might be isolated from the measuring electrode on passage of a bubble, is needed. Pulsations in the flow, arising in the peristaltic pump, were insufficient to cause significant pressure fluctuations at the probe membranes; such fluctuations can lead to cyclic variations in the thickness of the thin film, and the concomitant to and fro interchange between the thin film and bulk electrolyte solutions results in oscillation of the probe e.m.f. The wash solution contained the determinant at a concentration about half that in the lowest concentration sample; this is preferable to using a distilled water wash, serving to raise the baseline and effectively reduce the response time. A similar system has been successfully applied[17] to the determination of ammonia in the digests from Kjeldahl nitrogen analyses, using an ammonia-sensing membrane probe. It is usually important in such systems to ensure that the probe is not unduly exposed to undiluted reagent, as would happen in the absence of sample or wash solutions, in order to prevent osmotic drift.

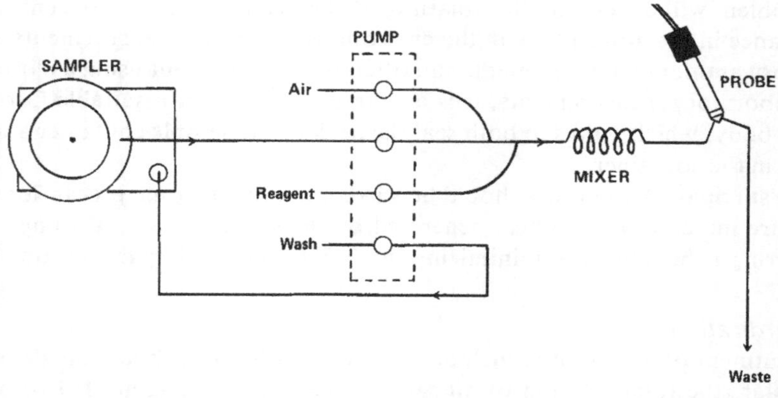

FIGURE 4. Diagram of continuous flow analytical system incorporating gas-sensing membrane probe.[7]

2. On-Line Monitoring

By comparison with laboratory systems for automatic analysis, there are several additional requirements which must be fulfilled by a satisfactory on-line monitor; in particular, it should be capable of long-term, preferably unattended, operation. The suitability of gas-sensing probes for on-line monitoring has been demonstrated by two applications of an ammonia-sensing membrane probe to the determination of ammonia in boiler feed-water.[18,19] In one,[18] an ammonia probe showed no loss of sensitivity or increase in response time over 4 months of continuous use, while a probe lifetime of 2 months was quoted from the results obtained in the other.[19] A further demonstration was the use of a carbon dioxide-sensing probe for the determination of carbon dioxide in power station waters,[14] where a probe was used continuously for 10 weeks without any deterioration in performance.

Virtually all the on-line monitors based on gas-sensing probes which are currently in use are for the determination of ammonia in water, although this situation can be expected to change in the future.

D. Gas Analysis

In principle, gas-sensing probes are very suitable for measurement of the partial pressures of gases in gas mixtures, providing that the sample is, or can be, saturated with water vapor to prevent loss of water from the internal electrolyte. In practice, however, there seem to be no known examples of their use for direct measurements in the gas phase—apart, of course, from "air-gap electrodes" themselves, which operate in this way. The reasons for this are not entirely clear, but in general it appears likely that gas-sensing probes would probably best meet the performance requirements for many gas analyses if used in on-line monitoring systems with facilities to control their environment, to pretreat and control the temperature and pressure of the sample, and to perform periodic automatic calibration; such on-line systems are relatively expensive, whereas a prime requirement is often for a low-cost on-line system.

V. SPECIFIC APPLICATIONS OF GAS-SENSING PROBES

A. Ammonia Probe

The ammonia-sensing probe is easily the most widely used of all the available gas-sensing probes, the major applications being in the analysis of fresh waters, effluents, and sewage, both treated and untreated.

Beckett and Wilson[12] used an ammonia-sensing membrane probe for the analysis of treated and untreated water from the River Thames and compared the results with those from an automated spectrophotometric method and a manual distillation/Nesslerization method, obtaining good agreement. The relative standard deviation for river water analysis was about 3% at the 1-mg/ℓ level of ammonia. Membranes were changed approximately every 3 months, and an alkaline reagent containing EDTA was used to prevent precipitation of metal hydroxides which might coat the probe membrane. A generally satisfactory pretreatment reagent, for addition at the rate of one volume per ten volumes of sample, comprises a 1 mol/ℓ sodium hydroxide solution containing 0.1-mol/ℓ EDTA. Evans and Partridge[20] used an ammonia-sensing membrane probe to analyze a wide range of samples, from potable water to crude sewage, for both total ammoniacal and albuminoid nitrogen, comparing the results with a spectrophotometric method and a Nessler method. Satisfactory agreement was obtained except for the albuminoid nitrogen determination at high ratios of free to albuminoid nitrogen. The precision of ammoniacal nitrogen determination was better than 4% for concentrations above 0.4 mg/ℓ. An alkaline reagent containing EDTA was again used, and chlorine-containing samples, such as swimming pool water, were treated with sodium thiosulfate solution prior to addition of the alkaline reagent to avoid interference from chloramines. Thomas and Booth[13] also used an ammonia-sensing membrane probe to analyze surface waters, saline waters, and sewage, finding that the probe method gave results which were as accurate as those from an automated spectrophotometric method but which were slightly more precise (standard deviation 0.038 mg/ℓ at a nitrogen concentration of 1 mg/ℓ for river water analysis). Similar findings have been reported by other authors.[15,21]

Midgley and Torrance[18] used an ammonia-sensing membrane probe in an on-line monitor to determine ammonia in boiler feed-water and condensate; with 12-hr automatic standardization, ammonia concentrations in the range of 0.1 to 1 mg/ℓ could be determined with a maximum relative standard deviation of 10%. Mertens et al.[19] also used an ammonia-sensing membrane probe in an on-line monitor for the same purpose, comparing the probe with an ammonium ion-selective electrode used under the same conditions; both gave results which agreed well with spectrophotometric methods (accuracy about 2%), but the ammonium electrode had a lifetime of only 2 weeks.

The determination of ammonia in the digests resulting from Kjeldahl total nitrogen determinations is an important application of the ammonia probe, as it allows the usual distillation step for removal of ammonia from the digest to be eliminated, thus reducing the overall analysis time considerably. The digests have very high osmotic strengths and are usually diluted before the alkaline reagent is added; even so, it is often necessary to modify the internal electrolyte of the probe to prevent osmotic drift. The reaction between the acidic digest and the alkaline reagent produces a significant quantity of heat, and some means of controlling sample temperatures is generally required. Catalysts such as copper or mercury, which may be present in the digest, must be removed by using a modified alkaline reagent including EDTA or iodide,[22] respectively. The use of the probe has been reported in the determination of the total nitrogen contents of meat products,[23,24] soils,[22] malt, wort and beer, fresh waters,[25] and waste water.[26] Closed flow systems are particularly advantageous in this application as the sample temperature can be closely controlled relatively easily; Buckee[17] and Stevens[25] used continuous flow systems to obtain rapid and accurate results.

The application of ammonia-sensing membrane probes to the determination of ammonia in blood has been reported by several workers;[27-30] an "air-gap electrode" has also been successfully applied.[8] Park and Fenton[28] used a simple continuous flow system to analyze plasma samples at the rate of 20 per hour. Addition of perchloric acid to the sample to precipitate proteins was found to be necessary by Proelss and Wright,[30] the sample then being centrifuged and the supernatant liquid analyzed; a limit of detection of 0.02 mg/ℓ was found for ammoniacal nitrogen.

Other applications include the determination of ammonia in airborne particles[31] and in tobacco smoke.[32] In the first case, the particles were filtered from the air and extracted with deionized water, while in the second, the ammonia was absorbed from the gas by scrubbing; the ammonia content of the resulting solution was then determined with an ammonia probe in each case.

The high selectivity of ammonia probes has led to attempts to apply them to the indirect determination of nitrate in water after reduction to ammonia; nitrate ion-selective electrodes have comparatively poor selectivity. Mertens et al.[33] studied this technique and concluded that Devarda alloy bonded into polystyrene or poly (vinyl chloride) beads was the most satisfactory reductant; the optimum pH for reduction was

13. A reduction column utilizing this material was incorporated, with an ammonia-sensing membrane probe, into a continuous flow analysis system which gave a linear calibration graph, for nitrate concentrations in the range 1 to 50 mg/ℓ, with a sampling rate of 20 per hour; for this concentration range, the reduction efficiency was 100% with a freshly prepared column, falling to 80% after 5 days but still giving a linear calibration graph. Results agreed well with those obtained by UV spectrophotometry, and relative standard deviations of 2.5 and 1.9% were found at nitrate levels of 3 and 30 mg/ℓ, respectively.

The major application of ammonia probes to indirect determinations is their use to detect the products of the reactions of enzymes with other substances; Moody and Thomas,[34] Guilbault,[35] and Bailey[6] have reviewed the use of gas-sensing probes for this purpose (see also Chapter 2, Volume II). Important contributions have been made by Rechnitz and co-workers[36,37] and Johansson and co-workers,[38,39] while commercial developments have been reviewed by Gray et al.[40]

B. Sulfur Dioxide Probe

Few applications of sulfur dioxide probes have been reported in the literature. Their main area of application is the determination of sulfur dioxide in foods and beverages, where they can generally provide simpler and more rapid analysis than the traditional methods; this area has been briefly reviewed by Barnett[41] and Bailey.[42]

Krueger[43] determined sulfur in petroleum products with a sulfur dioxide-sensing membrane probe; the fuel was burned in a sulfur lamp apparatus, and the sulfur dioxide produced was collected in a solution of tetrachloromercurate ions, containing glycerol to inhibit oxidation, at pH 6.9. This solution was then treated with sulfamic acid to remove nitrate ions and acidified, the sulfur dioxide being determined by a known addition procedure; the detection limit was below 0.005% sulfur. The same absorption solution is suitable for scrubbing sulfur dioxide from stack gases for subsequent measurement.

C. Carbon Dioxide Probe

Carbon dioxide probes were originally developed to measure the partial pressure of carbon dioxide in blood, and they continue to be very widely used for this purpose, quite often in automated systems for blood gas analysis. Apart from a few isolated examples, such as the determination of carbon dioxide in soil-root systems,[44] it is only relatively recently that other areas of application have been reported.

Midgley[14] used a carbon dioxide-sensing membrane probe in an on-line monitor to analyze power station waters, obtaining good performance, although the limit of detection of about 0.1 mg/ℓ was too high for the system to be used routinely for continuous monitoring. The probe was also used for discrete sample analysis in a laboratory continuous flow system; standard deviations ranged from 0.13 to 2.0 mg/ℓ at concentrations of 2.32 and 23.2 mg/ℓ, respectively.

Fiedler et al.[45] applied the carbon dioxide-sensing "air-gap electrode" to the determination of total inorganic carbon and total organic carbon contents of water. Total inorganic carbon was determined after simple acidification of the sample, while total organic carbon was determined by oxidation to carbonate, by boiling with alkaline potassium persulfate solution under pressure, and acidification. Satisfactory results were obtained for carbonate concentrations above 2×10^{-3} mol/ℓ but preconcentration was required for lower concentrations because of the restricted Nernstian range of the sensor. Up to 100-fold preconcentration was possible by a precipitation/dissolution technique involving lead carbonate, and excellent recoveries were obtained for samples containing from 5×10^{-3} to 10^{-4} mol/ℓ carbonate after allowing for the effect of atmospheric carbon dioxide.

Carbon dioxide probes have been used for indirect determinations involving enzymes; for example, urea and tyrosine have been determined[46] by reaction with the appropriate decarboxylase coated, as a suspension, onto the probe membrane.

D. Nitrogen Oxide Probe

Nitrogen oxide probes have been applied to the determination of nitrite in water, soil extracts,[47] and smoked fish.[48] The somewhat ambiguous name of the probe arises because the species sensed is the mixture of nitrogen oxides and nitrous acid generated when a nitrite solution is acidified; thus, nitrite is considered to be the determinant.

Tabatabai[47] used a nitrogen oxide-sensing membrane probe to determine nitrite in water samples and soil extracts, obtaining results which agreed well with those from a modified Griess-Ilosvay method. The method covered the concentration range 0.1 to 5 mg/ℓ of nitrite nitrogen with good precision, relative standard deviations generally being less than 1%.

Sherken[48] used a known addition method to determine nitrite in aqueous extracts of smoked fish with a nitrogen oxide-sensing membrane probe. Relative standard deviations of 1.6 and 6.3% were obtained at the highest and lowest concentrations tested, respectively, with good recoveries in the range 200 to 2000 mg/ℓ nitrite in the extract.

E. Other Probes

Gas-sensing probes for two other determinants have been reported recently. Hsiung et al.[49] used an "air-gap electrode" with an electrolyte solution comprising 5×10^{-3} mol/ℓ methylamine and 5×10^{-2} mol/ℓ potassium chloride to make a methylamine sensor. The performance of the resulting sensor was not encouraging; response was very slow, linear but non-Nernstian over only about 2 decades of concentration, and dimethylamine and ammonia interfered strongly. In the linear response region, the relative standard deviation was about 1.4% when the measurement time was 24 min, but was about 5% for a measurement time of 12 min.

Chang et al.[50] replaced the internal electrolyte solution of an ammonia-sensing membrane probe with a solution comprising 0.01 mol/ℓ trimethylamine hydrochloride and 0.04 mol/ℓ potassium chloride to make a trimethylamine-sensing probe which was used to analyze aqueous extracts of fish, comparing the results with an absorptiometric method. The response to trimethylamine was fast, linear, and almost theoretical between 10^{-2} and 10^{-4} mol/ℓ but significant responses to ammonia, methylamine, and dimethylamine were also observed; in fact, when dimethylamine was substituted for trimethylamine in the internal electrolyte, the probe exhibited a linear response to dimethylamine. Correlation between the two methods was satisfactory for standard solutions but much poorer for fish extracts, probably due to the poor selectivity. Despite this, it was considered that the probe method was so simple and economical that it was suitable for quality control purposes.

REFERENCES

1. Stow, R. W., Baer, R. F., and Randall, B. F., *Arch. Phys. Med. Rehabil.*, 38, 646, 1957.
2. Severinghaus, J. W. and Bradley, A. F., *J. Appl. Physiol.*, 13, 515, 1957.
3. Severinghaus, J. W., *Ann. N.Y. Acad. Sci.*, 148, 115, 1968.
4. Smith, A. C. and Hahn, C. E. W., *Br. J. Anaesth.*, 41, 731, 1969.
5. Ross, J. W., Riseman, J. H., and Krueger, J. A., *Pure Appl. Chem.*, 36, 473, 1973.

6. Bailey, P. L., *Analysis with Ion-Selective Electrodes*, Heyden, London, 1976, chap. 7.
7. Bailey, P. L. and Riley, M., *Analyst (London)*, 100, 145, 1975.
8. Růžička, J. and Hansen, E. H., *Anal. Chim. Acta*, 69, 129, 1974.
9. Midgley, D. and Torrance, K., *Analyst (London)*, 97, 626, 1972.
10. Hansen, E. H. and Larsen, N. R., *Anal. Chim. Acta*, 78, 459, 1975.
11. Bailey, P. L. and Riley, M., *Analyst (London)*, 102, 213, 1977.
12. Beckett, M. J. and Wilson, A. L., *Water Res.*, 8, 333, 1974.
13. Thomas, R. F. and Booth, R. L., *Environ. Sci. Technol.*, 7, 523, 1973.
14. Midgley, D., *Analyst (London)*, 100, 386, 1975.
15. Gilbert, T. R. and Clay, A. M., *Anal. Chem.*, 45, 1757, 1973.
16. Analytical Methods Guide, 7th ed., Orion Research Inc., Cambridge, Mass., 1975.
17. Buckee, G. K., *J. Inst. Brew. London*, 80, 291, 1974.
18. Midgley, D. and Torrance, K., *Analyst (London)*, 98, 217, 1973.
19. Mertens, J., Van den Winkel, P., and Massart, D. L., *Bull. Soc. Chim. Belg.*, 83, 19, 1974.
20. Evans, W. H. and Partridge, B. F., *Analyst (London)*, 99, 367, 1974.
21. Banwart, W. L., Tabatabai, M. A., and Bremner, J. M., *Commun. Soil Sci. Plant Anal.*, 3, 449, 1972.
22. Bremner, J. M. and Tabatabai, M. A., *Commun. Soil Sci. Plant Anal.*, 3, 159, 1972.
23. Todd, P. M., *J. Sci. Food Agric.*, 24, 488, 1973.
24. Deschreider, A. R. and Meaux, R., *Analusis*, 2, 442, 1973.
25. Stevens, R. J., *Water Res.*, 10, 171, 1976.
26. McKenzie, L. R. and Young, P. N. W., *Analyst (London)*, 100, 620, 1975.
27. Coleman, R. L., *Clin. Chem. (Winston-Salem, N.C.)*, 18, 867, 1972.
28. Park, N. J. and Fenton, J. C. B., *J. Clin. Pathol.*, 26, 802, 1973.
29. Sanders, G. T. B. and Thornton, W., *Clin. Chim. Acta*, 46, 465, 1973.
30. Proelss, H. F. and Wright, B. W., *Clin. Chem. (Winston-Salem, N.C.)*, 19, 1162, 1973.
31. Fagan, M. L. and Dubois, L., *Anal. Chim. Acta*, 70, 157, 1974.
32. Sloan, C. P., and Morie, G. P., *Anal. Chim. Acta*, 69, 243, 1974.
33. Mertens, J., Van den Winkel, P., and Massart, D. L., *Anal. Chem.*, 47, 522, 1975.
34. Moody, G. J. and Thomas, J. D. R., *Analyst (London)*, 100, 609, 1975.
35. Guilbault, G. G., *Bull. Soc. Chim. Belg.*, 84, 679, 1975.
36. Llenado, R. A. and Rechnitz, G. A., *Anal. Chem.*, 74, 1109, 1974.
37. Thompson, H. and Rechnitz, G. A., *Anal. Chem.*, 74, 246, 1974.
38. Johansson, G. and Ögren, L., *Anal. Chim. Acta*, 84, 23, 1976.
39. Johansson, G., Edstrom, K., and Ögren, L., *Anal. Chim. Acta*, 85, 55, 1976.
40. Gray, D. N., Keyes, M. H., and Watson, B., *Anal. Chem.*, 49, 1067A, 1977.
41. Barnett, D., *CSIRO Food Res. Q.*, 35, 68, 1975.
42. Bailey, P. L., *J. Sci. Food Agric.*, 26, 558, 1975.
43. Krueger, J. A., *Anal. Chem.*, 46, 1338, 1974.
44. Jensen, C. R., Van Grundy, S. D., and Stolzy, L. H., *Soil Sci.. Soc. Proc.*, 29, 631, 1965.
45. Fiedler, U., Hansen, E. H., and Růžička, J., *Anal. Chim. Acta*, 74, 423, 1975.
46. Guilbault, G. G. and Shu, F. R., *Anal. Chem.*, 44, 2161, 1972.
47. Tabatabai, M. A., *Commun. Soil Sci. Plant Anal.*, 5, 569, 1974.
48. Sherken, S., *J. Assoc. Off. Anal. Chem.*, 59, 971, 1976.
49. Hsiung, K. P., Kuan, S. S., and Guilbault, G. G., *Anal. Chim. Acta*, 84, 15, 1976.
50. Chang, G. W., Chang, W. L., and Lew, K. B. K., *J. Food Sci.*, 41, 723, 1976.

Chapter 2

ENZYME ELECTRODES

P. Vadgama

TABLE OF CONTENTS

I. INTRODUCTION

Following the widespread exploitation of ion-selective electrodes in the measurement of ionized species, the last decade has also seen an increasing interest in the use of these electrodes in the measurement of organic species. The organic molecule is made to undergo an enzyme-catalyzed reaction with either consumption or generation of an ion for which an ion-selective electrode is available. Usually the enzyme is held in a membrane layer over the sensor surface of the electrode, and the two constitute a combination which is usually known as an enzyme electrode. Other types of enzyme electrodes have been constructed in which the enzyme is used in conjunction with an amperometric device, but in this chapter, only potentiometric devices will be described.

Yet another kind of enzyme electrode is where substrate and electrode are used together to measure enzyme activity, but such electrodes have received little attention so far (Section V). The latest development is to use the enzyme while still present (unisolated) in a living organism (Section VI).

Enzyme electrodes permit measurement of organic molecules with the same facility that ion-selective electrodes permit the measurement of ions. Thus, analysis requires little manipulation of the sample, and rapid, highly selective measurements are possible which can be made continuous and amenable to automation. The use of an enzyme as the catalyst permits reactions to take place under mild conditions of temperature and pH and at very low substrate concentration. Furthermore, with the correct choice of enzyme, reactions can be made very specific, thereby enabling enzyme electrodes to be highly selective for a given substrate.

II. THEORY

The response characteristics of enzyme electrodes are subject to many variables in their design and construction, and an understanding of the theoretical basis of their function would help to improve their performance. Initial work considered only steady-state behavior,[1,2] but more recently computer-based numerical methods have been used to examine both steady-state and transient behavior.[3]

For the simplest case where a single enzyme is used to transform a simple substrate (S) into a single product (P) measured at the interface between the enzyme layer and electrode surface, the usual equation can be used to examine the steady state:

$$E + S \underset{K_{-1}}{\overset{k_{+1}}{\rightleftharpoons}} ES \xrightarrow{k_{+2}} E + P \tag{1}$$

where E is the free enzyme.

From Michaelis-Menten kinetics, the rate of reaction

$$V = \frac{d\,[P]}{dt} = \frac{-d\,[S]}{dt} = V_m \frac{[S]}{K_m + [S]} \tag{2}$$

where V_m is the maximum rate of reaction, and K_m is the Michaelis constant for the immobilized enzyme.

Within the enzyme layer, [S] and [P] are affected by diffusion as well as by the enzyme reaction. If it is assumed that the radius of curvature of the probe is very large as compared to the thickness of the enzyme layer, and that only one-dimensional diffusion occurs through the enzyme layer, then by Fick's Second Law:

$$\frac{d\,[P]}{dt} = D_P \frac{d^2\,[P]}{dx^2} + \frac{V_m\,[S]}{K_m + [S]} \tag{3}$$

where D_P is the diffusion constant for the product in the enzyme layer, and x is the distance from the outer surface of the enzyme layer.

For the steady-state, therefore,

$$D_P \frac{d^2\,[P]}{dx^2} = -\frac{V_m\,[S]}{K_m + [S]} \tag{4}$$

which for $[S] \ll K_m$ becomes

$$D_P \frac{d^2\,[P]}{dx^2} = -\frac{V_m}{K_m}\,[S] \tag{5}$$

Equation 5 can be integrated for a membrane of given thickness to obtain an equation describing the distribution of product concentration within that membrane. If the resulting equation is differentiated with respect to x, then evaluating the concentration gradients at the two membrane surfaces gives expressions for product fluxes which can be expressed solely in terms of bulk solution concentrations. Blaedel et al.[2] have derived such equations and, from these, have obtained an equation defining the steady-state concentration of product at the surface of the electrode sensor. Their model took account of the effects of external mass transfer and the partitioning of substrate and product between the bulk and enzyme phases. If such factors are disregarded, then a simpler expression can be obtained:[4]

$$[P]_{sensor} = [P]_{bulk} + \frac{D_S}{D_P} \frac{\cosh \alpha L - 1}{\cosh \alpha L} [S]_{bulk} \qquad (6)$$

where L is membrane thickness, D_s is the diffusion constant of the substrate, and

$$\alpha^2 = \frac{V_m}{K_m D_s}$$

Thus, $[P]_{sensor}$ is proportional to the bulk substrate concentration, and the electrode potential (E) should be proportional to $[S]_{bulk}$ if there has been no product build-up in the bulk solution. Equation 6 also serves to show that $[P]_{sensor}$ depends on several constant factors: K_m, enzyme activity, thickness of the enzyme layer, and diffusion coefficients for product and substrate.

At very high levels of enzyme activity or a thick enzyme layer, Equation 6 simplifies to:

$$[P]_{sensor} = [P]_{bulk} + \frac{D_S}{D_P} \cdot [S]_{bulk} \qquad (7)$$

For $[S] \gg K_m$, an expression for [P] at the sensor surface can be similarly obtained:

$$[P]_{sensor} = [P]_{bulk} + \frac{V_m L^2}{2 D_P} \qquad (8)$$

Equation 8 indicates that the electrode response should be independent of bulk substrate concentration. Blaedel et al.[2] obtained some experimental support for the above expressions in their study of the urea electrode.

Tranh-Minh and Broun[3] used similar equations, amenable to computer evaluation, and were able to derive profiles for substrate and product concentrations in the enzyme layer before the attainment of steady-state conditions. Carr,[5] by Fourier analysis of the appropriate differential equations, determined the transient response behavior but assumed that both product and substrate had the same diffusion constant (D). For the analytically important case $[S] \ll K_m$, he found the electrode response to be determined by a dimensionless parameter: $K_m D/V_m L^2$. This parameter has a value of zero for an ideal, diffusion-controlled, enzyme electrode with infinite enzyme activity. With the parameter set at 1, to simulate a suboptimal activity, the steady-state response decreased but the transient response was only slightly prolonged. This result indicated that with a steady-state response within acceptable limits, the response time is diffusion-controlled and can be little altered by addition of more enzyme. Of importance also was another dimensionless term: Dt/L^2. When this term becomes greater than 1.42, the electrode response was within 1 mV of its steady-state value, provided enzyme activity was acceptable. He concludes that long response times are to be expected for species whose diffusion coefficients are small or decreased by the enzyme membrane system.

III. THE ENZYME LAYER

The use of enzymes as reagents confers certain advantages which have already been commented upon. There are also problems associated with their use: they are expensive, not always sufficiently pure, relatively unstable, and tend to have a variable activity. Most of these problems can be overcome by using enzymes with an electrode rather than simply in solution with the substrate. Frequently, the enzyme is used in immobilized form in a layer over the electrode sensor[6] and can be either chemically immobilized or physically immobilized. In chemical immobilization, the enzyme is attached by means of a covalent bond to some immobilizing group, and in physical immobilization, the enzyme is trapped within an inert matrix such as starch or polyacrylamide gel, or just adsorbed onto some water-insoluble support. The mode of immobilization does have a functional significance since with chemical immobilization the enzyme is accessible to molecules of all sizes, whereas a physically trapped enzyme is isolated from large molecules. In any case, immobilization has a stabilizing effect on enzyme activity, particularly chemical immobilization.

An important drawback of chemical immobilization, however, is the potential damaging effect of immobilizing groups on the active site of the enzyme, although blocking of the active site can occur with physical immobilization. Furthermore, chemically immobilizing an enzyme is inconvenient and gives enzyme layers that are difficult to reproduce in activity and thickness. Another approach to immobilization has, therefore, been that of cross-linking the enzyme with an inert protein using a bifunctional agent such as glutaraldehyde.[7] The resulting membrane can then be applied to an electrode surface. Such membranes are easy to produce, using mild conditions, and give a predictable enzyme activity.

When a soluble enzyme is used and held as a thin film over the electrode surface by a dialysis membrane, the resulting enzyme electrode shows reduced stability and necessitates the use of a large excess of enzyme.[6] Electrodes with immobilized enzymes are relatively free of such problems and can show an actual increase in activity for the first 20 to 40 days. This is possibly the result of diffusion channels forming in the enzyme layer or of changes in conformation of the enzyme molecules themselves such that the proportion of active enzyme increases. Practical aspects of enzyme immobilization and electrode preparation have been reviewed by Guilbault.[7]

IV. DESCRIPTION AND DISCUSSION OF THE PRINCIPAL SYSTEMS

A. Urea Electrodes

This electrode system is the most studied so far. The first such electrode was reported by Guilbault and Montalvo[8] and comprised a glass ammonium ion-selective electrode over the surface of which was held a layer of physically immobilized urease. The diffusion of urea from the bulk solution resulted in the generation of NH_4^+ ions within the enzyme layer:

$$CO(NH_2)_2 + 2H_2O \xrightarrow{\text{urease}} 2NH_4^+ + 2HCO_3^- \qquad (9)$$

These NH_4^+ ions were sensed at the glass electrode surface, and because of the stoichiometric nature of the reaction, the urea concentration can be obtained.

The electrode was, in fact, initially employed as a sensitized electrode for NH_4^+ ion measurement.[9] Thus, the enzyme-coated glass electrode was found to have an identical response (56 mV/decade) to that of the uncoated electrode (Beckman® 39137) but shifted 31 mV more positive, and its linear response to NH_4^+ in NH_4Cl solutions was extended from 10^{-3} mol/ℓ down to 5×10^{-5} mol/ℓ. Pretreatment of the coated electrode with various solutions had no effect on this linear range. The extended range of the coated electrode was thought to be due to the enzyme acting as a cation exchanger.

The dynamic response of the urease-coated electrode to NH_4^+ indicated that response was diffusion controlled and that the speed of response was mainly determined by the thickness of the enzyme layer. For a 350-μm enzyme gel layer, 98% of the steady state was attained in 42 sec. As expected, the washout time (time to remove product build-up in the membrane) was found to be reduced with a decrease in gel layer thickness and in NH_4^+ concentration.

The electrode also had an enhanced sensitivity to other ions, although with the same selectivity order as the uncoated electrode:

$$Ag^+ > K^+ > H^+ > NH_4^+ > Na^+ > Li^+ >> Mg^{++} = Ca^{++}$$

This led to the interesting conclusion that either the selectivity of the enzyme layer was similar to that of the cation electrode or that a high permeability to ions in the gel layer masked the selectivity of the enzyme exchanger.

For the measurement of urea, Guilbault and Montalvo constructed three types of electrodes.[10] Type I was made by covering a Beckman® 39137 glass electrode with acrylamide gel-immobilized urease, for type II a cellophane film was added over the enzyme layer, and for type III the enzyme gel was sandwiched between two cellophane films. The response of all the elctrodes increased with increasing urease concentration in the gel layer up to 20 mg urease per milliliter of gel. Further increase had little effect on response, and response was also independent of both the concentration of gel and its cross-linking. With a 350-μm gel layer and 175 mg of urease per milliliter of gel, all the electrodes performed similarly, the linear portions of the calibration curves being from 5×10^{-5} to 10^{-3} mol/ℓ urea. As predicted by theory, the response became independent of urea concentration with larger amounts of urea in solution, and for very low urea concentrations, the response tailed off because of the poor response of the glass electrode. The response times of the electrode system showed strong dependence on gel layer thickness; 98% of the steady-state response was obtained in 26 sec with a 60-μm layer and in 59 sec with a 350-μm layer. The presence of cellophane coatings had little effect on these response times. After optimizing the immobilization conditions of the urease, a type I electrode was produced which was stable for 14 days. This stability was extended to 21 days for the type II and III electrodes, and this was thought to be due to the presence of the outer cellophane layer preventing any leaching out of the urease.

Because of Na^+ and K^+ interference and the possible liquid junction effects with biological fluids, Guilbault and Hrabankova[11] added ion-exchange resin to samples to remove Na^+ and K^+. They also tried an uncoated NH_4^+ ion glass electrode as the reference electrode, but similar calibration curves were obtained. Since the resins used tended to lower sample pH and affect the electrode response, the best results were obtained when a small amount of resin (Dowex® 50W-X2) was added directly to samples (1 to 2 g/50 mℓ solution) and the electrode response measured after stirring. An alternative method of adding large amounts of resin, with subsequent dilution of sample in Tris buffer, gave reproducible results. As much as 5×10^{-3} mol/ℓ Na^+ and 10^{-4} mol/ℓ K^+ were found not to interfere with the response to 10^{-5} mol/ℓ urea solutions. With blood and urine samples, an accuracy of 2 to 3% was obtained. The electrode required daily standardization since the calibration curve could shift several millivolts while the slope remained unchanged.

An improved selectivity was also obtained by the use of an NH_4^+ ion-selective electrode based on the antibiotic nonactin.[12] The nonactin was embedded in a silicone rubber matrix forming the selective membrane (0.2 mm thick). A silver foil in 0.1 mol/ℓ NH_4Cl comprised the internal reference electrode. The selectivity coefficients for this nonactin electrode were $k_{NH_4,K}$ = 0.15, $k_{NH_4,K}$ = 1.3 × 10^{-4}, and much smaller for other ions. Selectivity was further enhanced for urea since two NH_4^+ ions were formed for every molecule of urea hydrolyzed. Also, since K^+ ion concentrations in plasma only vary in the range 3.5 to 5.0 mmol/ℓ, such changes had little effect on the urea results. Using an uncoated nonactin electrode as the reference electrode, K^+ ion interference was virtually eliminated. Electrode activity remained stable for over 1 week, and response times between 60 and 180 sec were obtained, depending on the gel layer. Electrode stability was subsequently enhanced by the use of chemically immobilized urease in the gel layer.[13] The urease was coupled to a cross-linked polymer, based on acrylic acid, which bore diazotized $-NH_2$ groups. The nature of such a gel could be varied; a gel with good capacity to couple with the enzyme had poor physical stability, and the best gel was obtained where there were six cross linkages to every coupling site. Electrodes so constructed retained their activity for over a month. One further refinement was the use of an uncoated electrode with a calomel reference electrode to effect dilution of samples to a constant K^+ interference level prior to measurements with the enzyme electrode.

A gas-sensing NH_3 electrode with a hydrophobic membrane is free from cation interference. Such an electrode (Orion® 95-10) was used by Rogers and Pool[14] to estimate urea in raw sewage samples. Soluble urease was first used to liberate NH_4^+ and then the sample made alkaline (pH 12) to generate NH_3, which was sensed at the ammonia electrode. Anfält et al.[15] succeeded in immobilizing urease directly onto the gas-permeable membrane of the electrode using glutaraldehyde as a cross-linking agent. Urease has an optimum activity at pH 6.5, but by using pH 7 to 8, sufficient NH_3 was produced to permit urea measurement, although half of the enzyme activity was lost in this pH range. Close agreement was obtained with theoretical calibration curves where complete hydrolysis of urea in the membrane phase was assumed. Response times of 1.5 to 2 min were obtained for urea between 10^{-3} to 10^{-2} mol/ℓ. The authors concluded that any further improvement in response time would be limited by the dynamic response of the ammonia electrode itself. A deterioration in the response time was noted after 20 days when biological fluids were used, clogging of the pores of the electrode membrane being judged as important.[16]

The Orion® 95-10 ammonia electrode has been used with EDTA-stabilized urease for whole blood measurements;[17] the urease solution was trapped between the electrode membrane and a cellophane dialysis membrane. Such an electrode permitted urea measurement with minimal pretreatment of the samples, but with blood the response time was 6 to 8 min. The electrode gave a stable response for 3 weeks, after determination of 150 samples.

The response times of these electrodes were considered by Mascini and Guilbault to be too long.[18] They used a thin Teflon® gas-permeable membrane (10 to 35 μm thick) in conjunction with a Radiometer E 5036/0 electrode and obtained improved responses. The urease was chemically immobilized over the Teflon® with glutaraldehyde and albumin and formed a thin, highly active layer. The response time was found to be half that of the Orion® electrode, and the electrode was stable for over 2 months (1000 assays).

Ruzicka and Hansen[19] showed that an air-gap electrode for NH_3 offered certain advantages over one employing a gas-permeable membrane. By avoiding the membrane construction and utilizing a very thin layer of electrolyte on the surface of the indicator electrode, a higher speed of response could be obtained; furthermore, the lifetime of the sensor was increased since there was no longer any direct contact with the samples, the gaseous NH_3 liberated from alkaline samples being sensed at the NH_4Cl-covered indicator electrode.

Immobilized urease was used in the sample chamber of this type of electrode in serum urea determinations.[20] At pH 8.5, sufficient enzyme activity was retained to liberate a measureable quantity of NH_3, and for a 100-μm enzyme layer, a response time of 2 to 4 min was obtained. Also of interest was the rapid return to the base line (within 30 sec). The electrode system was robust and could be used for up to a month (500 samples), giving a precision and accuracy of less than 2%. Later, the urease was immobilized directly onto the surface of the stirrer used to mix the samples in the electrode chamber. This allowed simultaneous mixing and hydrolysis of the urea, and the electrode response became quite fast (2 min for 10^{-4} mol/ℓ urea).

Recently a micro air-gap NH_3-sensing probe has been fashioned as a urea sensor.[21] The device comprised a glass micropipette with a tip diameter of less than 10 μm, on the inner wall of which was a layer of chemically immobilized urease. The electrode had a response time of several minutes and showed poor sensitivity, but this could be optimized.

Continuous flow techniques have also been used with NH_3 sensors to measure urea concentration. Thus, Johansson and Ögren[22] have used an Electronic Instruments Ltd. (Chertsey, Surrey, U.K.) 8002-2 NH_3 gas electrode in conjunction with immobilized urease in a reaction chamber. The urea samples are pumped through the chamber, and the urea is quantitatively converted to NH_4^+. Subsequent mixing with NaOH permitted the electrode sensor to operate at its optimum pH. Temperature, flow rate, and enzyme activity were found not to affect measurements, and the system could be calibrated with either urea or NH_4Cl solutions. Changes in ionic strength of the solutions affected response, probably through its effect on the activity of NH_3. The response time was 3 min.

A urea analyzer based on these principles has now become commercially available (Owens - Illinois Inc., Toledo, Ohio).[23] Here, the urease is immobilized on a porous alumina support allowing a large amount of enzyme to be immobilized and enabling direct injection of the samples. A gas-sensing NH_3 electrode is employed, and again separation of electrode and enzyme function allows optimization of pH conditions.

Another approach to the measurement of urea has been to use a pH electrode to sense the change in hydrogen ion concentration accompanying urea hydrolysis with urease.[24] The response was slow, however, 10 min at 5×10^{-4} mol/ℓ urea, with a physically immobilized urease layer 400 μm in thickness. The change in pH was adequate at a buffer strength of up to 10^{-2} mol/ℓ. The electrode had a stable response over a 2-week period.

A CO_2 sensor was used by Guilbault and Shu[25] to measure the HCO_3^- produced from the urea hydrolysis. Such an electrode system, where the urease was applied to the electrode surface, was very specific for urea but had a sluggish response (1 to 3 min) compared with the uncoated electrode. The need to compensate for the CO_2 present in blood samples was an important disadvantage.

B. Glucose Electrodes

The first reported use of an enzyme electrode was for the determination of glucose.[26] This utilized glucose oxidase in a film between two Cuprophane® membranes. In the presence of glucose, oxygen was consumed:

$$\text{Glucose} + O_2 + H_2O \xrightarrow[\text{oxidase}]{\text{glucose}} H_2O_2 + \text{gluconic acid} \qquad (10)$$

and the change in pO_2 was detected using an amperometric oxygen electrode:

$$O_2 + 2H_2O + 4e \longrightarrow 4OH^- \quad (11)$$

Subsequently the production of H_2O_2 has been measured amperometrically:[27]

$$H_2O_2 \longrightarrow O_2 + 2H^+ + 2e \quad (12)$$

Such amperometric devices will not be further discussed.

Nagy et al.[28] described a potentiometric glucose electrode based on an I^- ion-selective electrode. In this electrode, the H_2O_2 liberated from the glucose oxidase-catalyzed reaction, oxidized I^- added to the sample solutions. Peroxidase was necessary for this reaction:

$$H_2O_2 + 2I^- + 2H^+ \xrightarrow{\text{Peroxidase}} I_2 + 2H_2O \quad (13)$$

The glucose oxidase and peroxidase were immobilized in an enzyme layer over the iodide electrode, and local changes in I^- concentration were detected at the sensor surface. The best response was obtained using glucose oxidase and peroxidase in the ratio 2:1. Whether the enzymes were physically or chemically immobilized did not affect the steady-state response. Increasing the enzyme activity, however, led to an increased response, and the best results were obtained using the enzymes in a liquid membrane. Stability, as expected, was greater for the immobilized enzyme electrode (30 days) than for the liquid membrane type electrode (40 hr). Response time varied from 2 to 8 min. Further, the electrode response could be increased by reducing the iodide ion concentration in the glucose samples. However, the usefulness of this was limited by the ultimate sensitivity of the iodide electrode.

Glucose oxidase is highly selective for β-D-glucose; however, interference was noted to occur from cellobiose and maltose and, to a lesser extent, from 2-deoxyglucose. More important for biological samples was the interference due to reducing substances such as uric acid, tyrosine, ascorbic acid, and Fe(II). These substances are able to decompose H_2O_2 and had to be destroyed by pretreatment of the samples. Thus, H_2O_2 and peroxidase were added to samples followed by catalase to destroy excess H_2O_2. The iodide electrode itself was free from chloride, phosphate, K^+, Na^+, Ca^{2+}, Mg^{2+}, and Fe^{3+} interferences.

A glucose electrode comprising glucose oxidase in a layer over a conventional pH electrode was constructed by Nilsson et al.[24] The pH change resulting from the generation of gluconic acid, in the presence of glucose, was measured by the electrode, a weak buffer being used. It was found necessary to saturate the samples with oxygen to ensure that it was present in nonrate-limiting concentration.

C. Amino Acid Electrodes

Enzyme electrodes for determining amino acid concentrations were first used by Guilbault and Hrabankova.[29] These workers used a monovalent cation glass electrode (Beckman® 39137) onto which had been applied a layer of L-aminoacid oxidase (L-AAO). L-Amino acid diffusing into this enzyme layer was decomposed to ammonium ion which was measured at the glass electrode surface:

$$RCHNH_3^+COO^- + H_2O + O_2 \xrightarrow{\text{L-AAO}} RCOCOO^- + NH_4^+ + H_2O_2 \quad (14)$$

The H_2O_2 further reacted nonenzymatically with the α-ketoacid:

$$RCOCOO^- + H_2O_2 \longrightarrow RCOO^- + CO_2 + H_2O \quad (15)$$

Immobilization of the amino acid oxidase within polyacrylamide gel gave rise to problems, since some decomposition of the enzyme occurred during the polymerization and also because the riboflavin catalyst in the gel solution was an inhibitor of the enzyme. The enzyme used (from moccasin venom) also tends to undergo a spontaneous inhibition[30] which is pH dependent but can be prevented by addition of chloride ion. Consequently, the electrodes needed to be stored in 10^{-2} mol/ℓ chloride solution at pH 5.5. Liquid membrane electrodes were found to have the greatest activity and were more stable (2 weeks). The calibration curves for the five amino acids studied had different slopes, since the activity of the enzyme varied with the amino acid. The order of response was found to be L-phenylalanine > L-leucine, L-methionine > L-alanine > L-proline. As might be expected, further increasing the L-AAO concentration in the enzyme layer did not improve performance after the maximum response had been attained but did result in increased stability. The optimum pH for the electrode response was found to correspond to the pH optimum (7 to 7.5) for the enzyme.

When a small amount of catalase was added to the enzyme layer, both electrode response and stability improved. The catalase destroyed H_2O_2, and the amino acid decomposition went to completion, the overall reaction being:

$$2RCHNH_3^+COO^- + O_2 \longrightarrow 2RCOCOO^- + 2NH_4^+ \quad (16)$$

An important drawback for biological samples was again the sensitivity of the glass electrode to Na^+ and K^+. This interference problem was reduced when a nonactin-based ammonium ion electrode was used.[31] Additionally, stability was improved by chemically immobilizing the enzyme in polyacryl gel. Response times varied between 60 and 180 sec.

An electrode based on measurement of iodide has also been constructed.[31] In this approach, potassium iodide was added to the samples, before analysis, and a double enzyme layer comprising L-AAO and horseradish peroxidase (HRP) was used with an iodide ion-selective electrode:

$$RCHNH_3^+COO^- + H_2O + O_2 \xrightarrow{\text{L-AAO}} RCOCOO^- + NH_4^+ + H_2O_2 \quad (17)$$

$$H_2O_2 + 2H^+ + I^- \xrightarrow{\text{HRP}} I_2 + 2H_2O \quad (18)$$

Local decrease in I^- was measured at the iodide electrode, and the enzymes were chemically immobilized. It was found that the rate of the electrode response (mV/min) showed a better linear relationship with amino acid concentration than did the steady-state response and that this was particularly so at high amino acid concentrations where the larger amounts of H_2O_2 produced were liable to inhibit the peroxidase-catalyzed reaction. The time for analysis was also reduced by the rate of response method (<30 sec).

The electrode was used to measure L-phenylalanine since other amino acids produced a smaller response. However, L-cysteine was found to give a response directly at the iodide electrode without enzyme layer. In order to get a totally specific electrode for L-phenylalanine, Hsuing et al.[32] used the enzyme L-phenylalanine ammonia lyase together with an air-gap electrode. The enzyme is very specific for L-phenylalanine and was used in homogeneous solution:

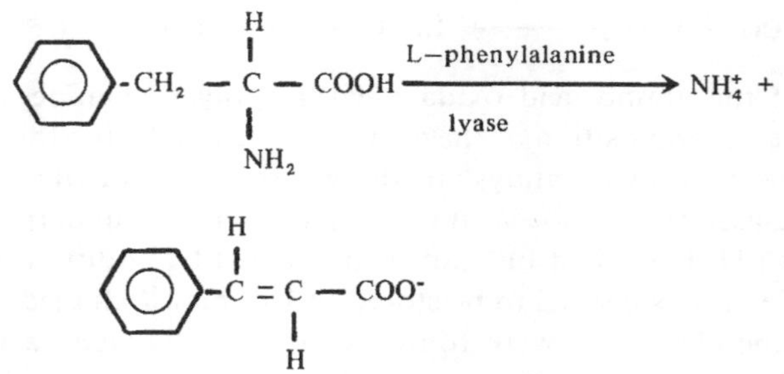

(19)

The NH_4^+ ion produced was detected as NH_3 after addition of alkali to the reaction mixture. There was no interference from either Na^+ and K^+ or other amino acids, but the low activity of the enzyme and a short linear range (10^{-4} to 6×10^{-4} mol/ℓ) proved to be disadvantageous.

D-Amino acids have also been measured with enzyme electrodes. Guilbault and Hrabankova[33] used an immobilized layer of D-amino acid oxidase (and also in solution) on a glass electrode (Beckman® 39137) and assayed D-phenylalanine, -alanine, -valine, -methionine, -leucine, -norleucine, and -isoleucine. All these amino acids produced a similar response with a linear range 10^{-4} to 5×10^{-2} mol/ℓ. The cyclic amino acid, D-proline, could not be measured since its deamination produced CH_3NH_2 instead of NH_4^+.

Although the electrodes using soluble enzyme gave a good initial response, they were far less stable than those using immobilized enzyme. When stored in flavin adenine dinucleotide (FAD), a cofactor for the enzyme, a stable electrode response could be obtained for over 21 days using the immobilized enzyme.

These same workers also measured L-asparagine using asparaginase:[33]

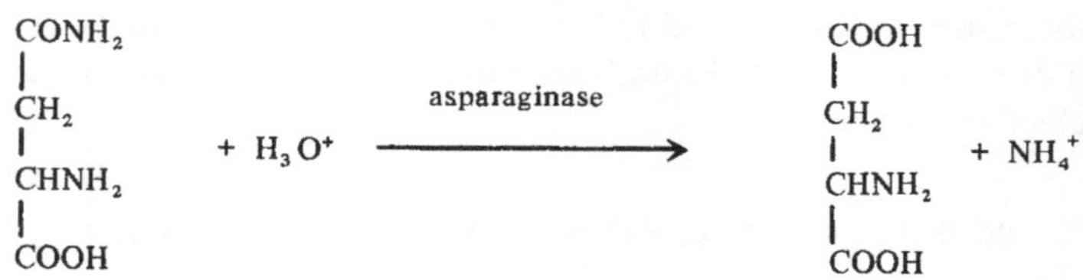

(20)

Immobilizing the enzyme extended electrode stability to over 21 days, and the enzyme did not require a cofactor.

L-Glutamine was measured by Guilbault and Shu[34] using the relatively stable enzyme glutaminase III (from microorganisms):

$$\text{Glutamine} \xrightarrow{\text{glutaminase}} \text{glutamate} + NH_4^+ \qquad (21)$$

There was no interference from other amino acids, although L-asparagine gave a small response, and the response time was 1 to 2 min. The pH optimum of the enzyme is 4.9, but since the glass electrode was sensitive to hydrogen ion, a compromise pH of 5.5 was used for the solutions. Immobilization of enzyme led to complete loss of activity, and so only liquid membrane electrodes could be produced, which were stable for only the first day.

The arginase-catalyzed hydrolysis of arginine liberates urea:

$$\text{Arginine} + H_2O \xrightarrow{\text{Arginase}} \text{Ornithine} + \text{urea} \qquad (22)$$

By incubating mixtures of arginase and urease with solutions of arginine, Neubecker and Rechnitz[35] measured the amount of NH_4^+ produced with a nonactin electrode and were able to estimate arginine. Although this is not a true enzyme electrode system, it does illustrate how an enzyme electrode might be useful in following reactions where an electroactive species is neither consumed nor generated. Furthermore, the pH optimum for arginase is 10, but that for urease is 6.5, so neither is ideal for the ion-selective electrode used. Despite this, a compromise pH of 8 permitted sensitive measurement.

An alternative enzyme electrode system was used by Guilbault and Shu[25] for tyrosine. Using the enzyme tyrosine decarboxylase

$$\text{Tyrosine} \xrightarrow[\text{decarboxylase}]{\text{tyrosine}} \text{tyramine} + CO_2 \qquad (23)$$

these workers monitored the liberation of CO_2. The enzyme was used as a liquid membrane held over a CO_2 sensor by a dialysis membrane. The electrode had a sluggish response, and although this could be improved by increasing pH, sensitivity was thereby impaired, as more of the CO_2 was converted to HCO_3^-. The electrode was stable for 3 weeks.

A carbonate ion-selective electrode has also been used for amino acids in conjunction with specific decarboxylases. Thus, Tong and Rechnitz[36] have measured L-lysine and L-arginine using the corresponding decarboxylases in homogeneous solution. It would seem feasible, therefore, to construct an enzyme electrode based on such a system.

Amperometric techniques have also been applied to amino acids. In one such method,[37] L-amino acid oxidase was coupled to a platinum electrode which sensed the peroxide produced:

$$H_2O_2 \longrightarrow O_2 + 2H^+ + 2e \qquad (24)$$

D. Amygdalin Electrode

This was the first enzyme electrode utilizing a nonglass membrane.[38] The electrode comprised a solid-state cyanide ion sensor on the surface of which had been coated a layer of β-glucosidase. The enzyme was physically immobilized in acrylamide gel and hydrolyzed amygdalin:

$$\begin{array}{l} C_6H_5CHCN \\ \quad | \\ \quad OC_{12}H_{21}O_{10} \end{array} \xrightarrow[\beta\text{-glucosidase}]{H_2O} 2C_6H_{12}O_6 + C_6H_5CHO + HCN \qquad (25)$$

At the high pH used (10 to 11), most of the HCN was ionized and enabled electrode measurements to be made.[39] The high sample pH, however, had an adverse effect on enzyme activity, and the electrode had a very slow response (30 min at 10^{-5} to 10^{-4} mol/ℓ). Furthermore, the enzyme leached out of the gel layer, thus shortening the electrode life to 4 days.

Mascini and Liberti[40] extended these studies using a liquid membrane electrode in which β-glucosidase was applied directly onto the electrode surface and covered with dialysis membrane. When this electrode was employed at pH 7, within the optimum pH range of the enzyme, a fast response time was obtained (1 to 2 min for 10^{-3} to 10^{-1} mol/ℓ amygdalin). With the diminished HCN ionization, however, the steady-state response was now decreased.

E. Penicillin Electrode

By immobilizing penicillinase (β-lactamase) in a polyacrylamide gel on a glass pH electrode, Papariello et al.[41] were able to estimate penicillin concentrations. The enzyme hydrolyzed penicillin, and the resulting penicilloic acid led to an increase in hydrogen ion concentration at the glass electrode surface:

$$\text{Penicillin} \xrightarrow[\text{penicillinase}]{H_2O} \text{Penicilloic acid} \qquad (26)$$

The best electrode response was obtained near the optimum pH for the enzyme (6.4) giving a response time of less than 30 sec. Penicillin is metabolized to penicilloic acid, and an important advantage of the electrode was that it did not respond to this metabolite. A similar electrode has been produced using penicillinase in a liquid film.[24]

Later work with the electrode, however, demonstrated that direct contact between penicillinase and the glass pH electrode led to the electrode becoming sensitive to other monovalent cations, with a reduced H^+ ion sensitivity. Adsorption of the enzyme onto the glass sensor surface was considered to be the explanation for this. Normal sensitivity was thus found to be restored if a dialysis membrane was interposed between the enzyme layer and the glass electrode surface. The resulting electrode was too sluggish to be of practical value, however, and a new kind of electrode system was devised in which the penicillinase was first adsorbed onto a fritted glass disk and the disk attached to a flat-form pH electrode. This electrode was free from cation interference. Response times of about 1 min were obtained with various penicillins at 10^{-4} mol/ℓ. At higher concentrations, response was much slower as the increased H^+ ion production diminished the activity of the enzyme. The electrode was stable over a 4-week period.

F. Creatinine Electrode

The enzyme creatininase has been employed in the construction of an electrode for creatinine.[43] The enzyme was present in a liquid membrane held around the sensor surface of an Orion® ammonia gas-sensing electrode (95-10). The enzyme converted creatinine to NH_3 and N-methylhydantoin, and the NH_3 was sensed at the electrode surface. The enzyme used required purification, which led to a much reduced activity. This was partly offset by the high optimum pH of the enzyme (8.5) at which the NH_4^+/NH_3 equilibrium was more favorable for the ammonia sensor. Enzyme activation with tripolyphosphate also served to increase electrode response. Nevertheless, a limited linear range was obtained (5.2×10^{-4} to 9.6×10^{-3} mol/ℓ), and response times varied between 2 and 10 min. There was no interference from creatine, urea, or arginine, and potentials were stable over 4 days.

G. Uric Acid Electrode

A CO_2 electrode has been used with immobilized uricase to measure uric acid concentration:[44]

$$\text{Uric acid} + O_2 \xrightarrow{\text{uricase}} \text{allantoin} + H_2O_2 + CO_2 \qquad (27)$$

The best results were obtained with the enzyme immobilized in cellulose, but response times varied between 5 and 10 min. Choice of buffer was found to be important; phosphate was an activator for uricase, and borate altered the stoichiometry of the reaction and reduced the amount of CO_2 produced. There was little change in electrode response over 10 days. The uricase reaction has also been used with an amperometric O_2 sensor for uric acid determinations.[45]

H. 5′-Adenosine Monophosphate Electrode

5′-adenylic acid deaminase selectively deaminates 5′-adenosine monophosphate (5′-AMP), and the stoichiometric production of NH_3 can be sensed at an NH_3 gas electrode:

$$5'\text{-AMP} \xrightarrow[\text{enzyme}]{H_2O} 5'\text{-inosine monophosphate} + NH_3 \qquad (28)$$

Papastathopoulos and Rechnitz[46] constructed an electrode for 5′-AMP using an Orion® 95-10 electrode on which a liquid film of 5′-AMP deaminase was confined by a cellophane membrane. The resulting electrode was stable over 4 days. The best results were obtained at pH 7.5, which was a compromise between the optimum pH for the enzyme and that which was ideal for the NH_3 sensor. Response times of about 2 min were obtained for concentrations near 10^{-2} mol/ℓ. When a cross-linked or adsorbed enzyme was employed, sensitivity was reduced, and the response became slow. There was no interference from 3′,5′-cyclic AMP, 5′-ATP, 5′-ADP, adenine, or adenosine.

Later, Riechel and Rechnitz[47] used the electrode with a low molecular weight cut-off membrane to study the binding between 5′-AMP and the enzyme D-fructose-1,6-diphosphatase. The modified electrode only responded to free 5′-AMP and enabled binding constants to be determined.

I. Lactate Dehydrogenase-Glutamate Dehydrogenase Electrode

Davies and Mosbach[48] constructed an electrode for glutamate to illustrate how the scope for enzyme electrodes might be increased by the use of immobilized coenzyme. They used an NH_4^+-sensitive glass electrode (Beckman® 39137) which was in contact with a liquid membrane. Included within the membrane were the enzymes glutamate dehydrogenase (GDH) and lactate dehydrogenase (LD) as well as their cofactor NAD^+ in a dextran-bound form. The cofactor was thus able to function within the liquid membrane without being leached out into the bulk solution.

With glutamate solutions, NH_4^+ was generated

$$\text{Glutamate} + NAD^+ \overset{GDH}{\rightleftharpoons} \alpha\text{-ketoglutarate} + NADH + H^+ + NH_4^+ \qquad (29)$$

and was sensed at the glass electrode. The solutions also contained excess pyruvate (2 mmol/ℓ) to effect the continuous regeneration of NAD^+:

$$\text{Pyruvate} + NADH + H^+ \overset{LD}{\rightleftharpoons} \text{Lactate} + NAD^+ \qquad (30)$$

It proved possible to measure pyruvate concentration, the samples now containing glutamate added in excess (10 mmol/ℓ), and the authors further considered measurement of α-ketoglutamate and L-lactate to be feasible since the enzymatic reactions are reversible.

V. SUBSTRATE ELECTRODES

A. Urease Electrode

Sensors have also been proposed for measuring enzyme activity. Montalvo[49] designed one for urease by reversing the principle of operation of the urea electrode. The sensor consisted of a Beckman® 39137 glass electrode which had been covered with cellophane membrane. Urea solution was able to pass in a thin film between the glass surface and the cellulose. When placed in urease samples, the urea diffused out from the electrode, and the NH_4^+ formed from its hydrolysis diffused back onto the sensor surface, thereby changing the potential.

In a later improved version,[50] a 150-μm nylon net under the cellophane membrane kept the urea space constant and a micro/roller pump maintained a constant flow of urea. The response of this electrode was linear for urea concentrations of up to 5 mol/ℓ. This much higher value as compared with the urea electrode indicates the considerable dilution of the urea that must occur in the urease samples. Above an optimal flow rate of the urea, the electrode response diminished, and this was probably due to the rate of removal of NH_4^+ in the urea layer approaching that of NH_4^+ diffusion across the layer. Stirring also reduced the response, since now more of the NH_4^+ was carried away by convection. The response to urea was linear in the range 6×10^{-4} to 10^{-1} urease unit per milliliter with a response time of 2 min.

B. Cholinesterase Electrode

Cholinesterase activity in serum was measured by Crochet and Montalvo[51] using a pH electrode in conjunction with a microelectrochemical cell. Two thin-layer solutions separated by a dialysis membrane comprised the cell. The sample was passed through the inner layer of this cell and an acetylcholine solution through the outer layer. The acetylcholine diffused into the enzyme layer with liberation of acetic acid, and the resulting change in H^+ ion was detected at the pH electrode:

$$\text{Acetylcholine} \xrightarrow{\text{Cholinesterase}} \text{Choline} + \text{acetic acid} \quad (31)$$

With a 10-μm thick enzyme layer, diffusion was found not to be rate limiting. Using the rate of the electrode response (mV/min), a linear calibration curve was obtained for cholinesterase activity from 10 to 70 Rappaport units per milliliter.

Guilbault[6] has referred to the use of an insoluble, Reineckate, salt of acetylcholine with a pH electrode:

$$\text{Acetylcholine Reineckate} \xrightarrow[\text{cholinesterase}]{} \text{acetic acid} + \text{choline - reineckate} \quad (32)$$

This procedure could be preferable to the use of the substrate in solution.

VI. BACTERIAL MEMBRANE ELECTRODES

Instead of combining isolated enzymes with an electrode sensor, a recent approach has been to utilize the enzyme while still present in a living microorganism. The activity of the enzyme can thus be optimized, and reactions requiring several enzymes and cofactors can become easily available. Thus, Rechnitz and co-workers[52] incorporated the microorganism *Streptococcus faecium* onto the surface of an NH_3 gas-sensing electrode (Orion Research Inc., Cambridge, Mass., 95-10) and were able to measure L-arginine, using the enzyme systems of the microorganism:

$$\text{L-Arginine} \xrightarrow[\text{deaminase}]{\text{arginine}} \text{citrulline} + NH_3 \quad (33)$$

$$\text{Citrulline} + H_3PO_4 \xrightarrow[\text{transcarbamylase}]{\text{ornithine}} \text{ornithine} + \text{carbamoylphosphate} \quad (34)$$

$$\text{Carbamoylphosphate} + \text{ADP} \xrightarrow[\text{kinase}]{\text{carbamate}} \text{carbamic acid} + \text{ATP}$$
$$\text{Carbamic acid} \longrightarrow CO_2 + NH_3 \tag{35}$$

Citrulline added to solutions did not interfere since it did not penetrate the bacterial cell walls. With a freshly prepared electrode, a linear range of 5×10^{-5} to 1×10^{-3} mol/ℓ arginine was obtained, but this decreased over 20 days. A major disadvantage was the long response time (20 min), which was probably due to diffusion through the cell walls. Also, the other enzyme systems in the microorganism caused interference from glutamine and asparagine.

The possibility of regenerating bacterial membrane electrodes was demonstrated by Kobos and Rechnitz[53] who measured L-aspartate using L-aspartase with an ammonia gas electrode:

$$\text{L–Aspartate} \xrightarrow{\text{L-aspartase}} \text{fumarate} + NH_3 \tag{36}$$

Although the enzyme was selective, it was very labile, and a conventional liquid membrane electrode only had 1 day of useful life. When a bacterial membrane electrode was used, based on a microorganism rich in the enzyme (*Bacterium cadaveris*), it was found that electrode function could be maintained by simply storing the electrode in a suitable growth medium at 30°C. Other bacterial enzyme systems, however, led to reduced specificity, and adenosine and asparagine also produced a response.

Rechnitz et al.[54] have similarly used the microorganism *Sarcina flava* to provide the unstable enzyme glutaminase and measured glutamine:

$$\text{Glutamine} \xrightarrow{\text{glutaminase}} \text{glutamate} + NH_3 \tag{37}$$

The response time was 5 min, and the electrode functioned well in solutions containing other amino acids, including plasma, where viscosity effects caused a shift in the calibration curve. The electrode was stable for 2 weeks.

VII. CONCLUSION

The survey of potentiometric enzyme electrodes (summarized in Table 1) given here shows the range of substances for which electrodes have been constructed so far. Their full potentiality has not yet been realized however, and with the increasing use of immobilized enzymes in analytical and preparative chemistry, the range must grow wider to permit simple, direct measurement of species which are difficult to measure using conventional analytical techniques. Also of interest is the prospect of continuous measurement with enzyme electrodes, and this may prove to be of considerable value in biology and medicine. Before these possibilities can be realized, however, numerous problems need to be overcome, particularly the lack of specificity and stability, but these will no doubt be amenable to optimization in the construction of these sensors.

TABLE 1

Summary of Potentiometric Enzyme Electrode Characteristics

Enzyme	Sensor electrode	Analyte	Linear range (mol/ℓ)	Slope (mV/decade)	Interference	Ref.
Urease	NH_4^+	Urea	5×10^{-5} to 10^{-3}	50	K^+, Na^+, H^+, Ag^+, Li^+	10
	NH_4^+	Urea	10^{-4} to 10^{-2}	55[a]	K^+, Na^+	12
	NH_3	Urea	10^{-3} to 10^{-2}	59.5—67.5	—	15
	NH_3	Urea	5×10^{-4} to 7×10^{-2}	90	—	17
	NH_3	Urea	5×10^{-3} to 10^{-2}	55	—	18
	NH_3	Urea	5×10^{-5} to 3×10^{-2}	53[a]	—	20
	NH_3	Urea	5×10^{-5} to 3×10^{-2}	57	—	22
	H^+	Urea	5×10^{-5} to 5×10^{-3}	41[a]	—	24
	CO_2	Urea	10^{-4} to 10^{-1}	57	Acetic acid	25
Glucose oxidase	H^+	Glucose	10^{-1} to 10^{-3}	50[a]	—	24
	I^-	Glucose	10^{-1} to 10^{-3}	160 (?)[a]	Maltose, cellobiose, reducing agents	28
L-AA Oxidase	NH_4^+	L-Amino acids	10^{-4} to 2×10^{-3}[a,b]	35[a,b]	Monovalent cations	29
	NH_4^+	L-Amino acids	2×10^{-4} to 10^{-2}[b]	35[a,b]	K^+, Na^+	31
L-AA oxidase and horseradish peroxidase	I^-	L-Amino acids	5×10^{-5} to 1×10^{-3}[b]	—	L-Cysteine (at iodide electrode)	31
L-Phenylalanine ammonia lyase	NH_3	L-Phenylalanine	10^{-4} to 6×10^{-4}	62[a]	—	32
Asparaginase	NH_4^+	L-Asparagine	10^{-4} to 10^{-3}[a]	45[a]	—	33
Glutaminase	NH_4^+	L-Glutamine	10^{-3} to 10^{-1}[a]	40—48	Monovalent cations and L-asparagine	34
Tyrosine decarboxylase	CO_2	L-Tyrosine	2.5×10^{-4} to 10^{-2}	55	Acetic acid	25
D-AA oxidase	NH_4^+	D-Amino acids	10^{-4} to 5×10^{-2}	30[a]	—	33

TABLE 1 (continued)

Summary of Potentiometric Enzyme Electrode Characteristics

Enzyme	Sensor electrode	Analyte	Linear range (mol/ℓ)	Slope (mV/decade)	Interference	Ref.
β-glucosidase	CN^-	Amygdalin	10^{-5} to 5×10^{-3}	48	Cu, Cd, Hg, and ions forming insoluble Ag salts or complexes with CN^-	39
	CN^-	Amygdalin	10^{-4} to 10^{-1}	53	—	40
Penicillinase	H^+	Penicillin	10^{-3} to 10^{-2}	83[a]	—	24
	H^+	Penicillin	10^{-4} to 5×10^{-2}	52	Monovalent cations	41, 42
	H^+	Penicillin	10^{-5} to 3×10^{-3}	56—58	—	42
Creatininase	NH_3	Creatinine	5.2×10^{-4} to 9.6×10^{-3}	53	—	43
Uricase	CO_2	Uric acid	10^{-4} to 2.5×10^{-3}	57	—	44
5′-AMP deaminase	NH_3	5′-AMP	8×10^{-5} to 1.5×10^{-2}	46	—	46
Glutamate dehydrogenase and lactate dehydrogenase	NH_4^+	Glutamate	10^{-4} to 10^{-3}	15—20	K^+, Na^+	48
		Pyruvate	6×10^{-5} to 8×10^{-4}			
Arginine deaminase, ornithine transcarbamylase, and carbamate kinase	NH_3	L-Arginine	5×10^{-5} to 10^{-3}	42	Glutamine and asparagine	52
L-Aspartase	NH_3	L-Aspartate	3×10^{-4} to 7×10^{-3}	45—50	Adenosine and asparagine	53
Glutaminase	NH_3	Glutamine	10^{-4} to 10^{-2}	48.5	—	54

[a] Estimated from calibration curves.
[b] Slope for L-phenylalanine.

REFERENCES

1. Racine, P. and Mindt, W., *Experientia Suppl.*, 18, 525, 1971.
2. Blaedel, W. J., Kissel, T. R., and Boguslaski, R. C., *Anal. Chem.*, 44, 2030, 1972.
3. Tranh-Minh, C. and Broun, G., *Anal. Chem.*, 47, 1359, 1975.
4. Bowers, L. D. and Carr, P. W., *Anal. Chem.*, 48, 545A, 1976.
5. Carr, P. W., *Anal. Chem.*, 49, 799, 1977.
6. Guilbault, G. G., *Methods in Enzymology*, Vol. 44, Mosbach, K., Ed., Academic Press, New York, 1976, 579.
7. Guilbault, G. G., *Immobilized Enzymes, Antigens, Antibodies and Peptides*, Vol. 1, Weetal, H. W., Ed., Marcel Dekker, New York, 1975, 293.
8. Guilbault, G. G. and Montalvo, J. G., *J. Am. Chem. Soc.*, 91, 2164, 1969.
9. Montalvo, J. G. and Guilbault, G. G., *Anal. Chem.*, 41, 1897, 1969.
10. Guilbault, G. G. and Montalvo, J. G., *J. Am. Chem. Soc.*, 92, 2533, 1970.
11. Guilbault, G. G. and Hrabankova, E., *Anal. Chim. Acta*, 52, 287, 1970.
12. Guilbault, G. G. and Nagy, G., *Anal. Chem.*, 45, 417, 1973.
13. Guilbault, G. G., Nagy, G., and Kuan, S. S., *Anal. Chim. Acta*, 67, 195, 1973.
14. Rogers, D. S. and Pool, K. H., *Anal. Lett.*, 6, 801, 1973.
15. Anfält, T., Granelli, A., and Jagner, D., *Anal. Lett.*, 6, 969, 1973.
16. Proelss, H. F. and Wright, B. W., *Clin. Chem.*, (Winston-Salem), 19, 1162, 1973.
17. Papastathopoulos, D. S. and Rechnitz, G. A., *Anal. Chim. Acta*, 79, 17, 1975.
18. Mascini, M. and Guilbault, G. G., *Anal. Chem.*, 49, 795, 1977.
19. Růžička, J. and Hansen, E. H., *Anal. Chim. Acta*, 69, 129, 1974.
20. Guilbault, G. G. and Tarp, M., *Anal. Chim. Acta*, 73, 355, 1974.
21. Paul, C. P., Rechnitz, G. A., and Miller, R. F., *Anal. Chem.*, 50, 330, 1978.
22. Johansson, G. and Ögren, L., *Anal. Chim. Acta*, 84, 23, 1976.
23. Gray, D. N., Keyes, M. H., and Watson, B., *Anal. Chem.*, 49, 1067A, 1977.
24. Nilsson, H., Åkerlund, A., and Mosbach, K., *Biochim. Biophys. Acta*, 320, 529, 1973.
25. Guilbault, G. G. and Shu, F., *Anal. Chem.*, 44, 2161, 1972.
26. Clark, C. C. and Lyons, C., *Ann. N.Y. Acad. Sci.*, 102, 29, 1962.
27. Guilbault, G. G. and Lubrano, G. J., *Anal. Chim. Acta*, 64, 439, 1973.
28. Nagy, G., von Storp, L. H., and Guilbault, G. G., *Anal. Chim. Acta*, 66, 443, 1973.
29. Guilbault, G. G. and Hrabankova, E., *Anal. Chem.*, 42, 1779, 1970.
30. Kearney, E. B. and Singer, T. P., *Arch. Biochem. Biophys.*, 33, 377, 1951.
31. Guilbault, G. G. and Nagy, G., *Anal. Lett.*, 6, 301, 1973.
32. Hsuing, C. P., Kuan, S. S., and Guilbault, G. G., *Anal. Chim. Acta*, 90, 45, 1977.
33. Guilbault, G. G. and Hrabankova, E., *Anal. Chim. Acta*, 56, 285, 1971.
34. Guilbault, G. G. and Shu, F. R., *Anal. Chim. Acta*, 56, 333, 1971.
35. Neubecker, T. A. and Rechnitz, G. A., *Anal. Lett.*, 5, 653, 1972.
36. Tong, S. L. and Rechnitz, G. A., *Anal. Lett.*, 9, 1, 1976.
37. Guilbault, G. G. and Lubrano, G. J., *Anal. Chim. Acta*, 69, 183, 1974.
38. Rechnitz, G. A. and Llendano, R. A., *Anal. Chem.*, 43, 283, 1971.
39. Llendano, R. A. and Rechnitz, G. A., *Anal. Chem.*, 43, 1457, 1971.
40. Mascini, M. and Liberti, A., *Anal. Chim. Acta*, 68, 177, 1974.
41. Papariello, G. J., Mukherji, A. K., and Shearer, C. M., *Anal. Chem.*, 45, 790, 1973.
42. Cullen, L. F., Rusling, J. F., Schleifer, A., and Papariello, G. J., *Anal. Chem.*, 46, 1955, 1974.
43. Meyerhoff, M. and Rechnitz, G. A., *Anal. Chim. Acta*, 85, 277, 1976.
44. Kawashima, T. and Rechnitz, G. A., *Anal. Chim. Acta*, 83, 9, 1976.
45. Nanjo, M. and Guilbault, G. G., *Anal. Chem.*, 46, 1769, 1974.
46. Papastathopoulos, D. S. and Rechnitz, G. A., *Anal. Chem.*, 48, 862, 1976.
47. Riechel, T. L. and Rechnitz, G. A., *Biochem. Biophys. Res. Commun.*, 74, 1377, 1977.
48. Davies, P. and Mosbach, K., *Biochim. Biophys. Acta*, 370, 329, 1974.
49. Montalvo, J. G., *Anal. Chem.*, 41, 2093, 1969.
50. Montalvo, J. G., *Anal. Biochem.*, 38, 359, 1970.
51. Crochet, K. L. and Montalvo, J. G., *Anal. Chim. Acta*, 66, 259, 1973.
52. Rechnitz, G. A., Kobos, R. K., Riechel, S. J., and Gebauer, C. R., *Anal. Chim. Acta*, 94, 357, 1977.
53. Kobos, R. K. and Rechnitz, G. A., *Anal. Lett.*, 10, 751, 1977.
54. Rechnitz, G. A., Riechel, T. L., Kobos, R. K., and Meyerhoff, M. E., *Science*, 199, 440, 1978.

Chapter 3

ION-SELECTIVE ELECTRODES IN MEDICINE AND MEDICAL RESEARCH

D. M. Band and T. Treasure

I. THE PLACE OF ION-SELECTIVE ELECTRODES IN CURRENT MEDICAL PRACTICE

For many years, writers have pointed out the range of possible applications of ion-selective electrodes in medical care. In spite of the proliferating literature on these devices and much of the technology being known for at least a decade, the promise remains largely unfulfilled. In clinical medicine, the pH electrode is, of course, used universally as part of the "blood gas analysis" equipment, but other ion-selective electrodes tend to remain in the hands of research workers or enthusiasts with a special interest.

We have recently tested our impression that electrode techniques are not as widespread in medical practice as might be supposed from reading the electrochemical literature. A questionnaire was sent to the 12 London Teaching Hospitals. Only one had an ion-selective electrode in routine clinical use. This was a fluoride electrode used by a renal unit in the assessment of patients undergoing renal dialysis. One hospital had an automatic analyzer under trial. This had been constructed in the hospital workshops, using commercially available ion-selective electrodes for the measurement of Na^+, K^+, HCO_3^-, urea, and glucose.[1] There was one calcium and one ammonium electrode in specialist laboratories in other hospitals. This return was obviously from a small selected sample and may only serve to show that the London Teaching Hospitals are lagging behind the rest of the world. However, the finding was in sharp contrast to the encouraging reports in the literature of multielectrolyte analyzers in clinical use. Dahms[2] reported an automated electrode system for Na^+, K^+, H^+, and Cl^- as long ago as 1967, and since then, other multichannel systems have been described.[3,4] Why have they not become widely used?

There seem to be two factors which might explain this failure of ion-selective electrodes to live up to their promise. The first is that some of the early electrodes that were marketed failed to fulfill the claims that were made of them. This is serious enough because the clinical chemist, having been let down once, is reluctant to spend more money and time on devices which he has come to view as inherently unreliable. This disappointment, added to the necessarily conservative attitude of the medical profession, has produced caution.

The second factor may be a failure to provide equipment fulfilling a particular clinical need. This is a more fundamental difficulty and cannot be lightly dismissed as "teething problems." In the routine laboratory, automated and semiautomated emission flame photometers produce a large number of accurate electrolyte measurements for day-to-day patient care. The systems are well tested, and considerable investments in money and personnel training have helped to consolidate their position.

The fundamental limitations of electrode determinations make it doubtful whether a multichannel electrode system could be made to meet the specifications of the best emission flame photometers. For example, the normal range for plasma sodium (Table 1) is between 138 and 151 mmol/ℓ. This represents only 2.4 mV change in a theoretical sodium electrode at 37°C. It is therefore unlikely that discrimination comparable with a flame photometer can be achieved, so attempts to introduce ion-selective electrodes as straight replacements for the flame photometer have little practical advantage, except perhaps in cost. Even this may be misleading, for, although the ion-selective electrode is basically a cheap sensor, the associated fluid transport, read-out, servicing, and time spent out of service could make it less attractive.

TABLE 1

Composition of Normal Plasma

Ion	Mean[a]	95% Limits[a]
Sodium	144.5	138—151
Potassium	4.3	3.4—5.2
Magnesium	0.85	0.65—1.05
Calcium	2.6	2.4—2.8
Chloride	100	101—111
Bicarbonate	24.9	21.3—28.5

[a] All values are in millimoles per liter concentration.

Adapted from Diem, K. and Lentner, C., *Scientific Tables*, 7th Ed., Ciba-Geigy, Basle, Switzerland and Macclesfield, U.K., 1970.

In spite of this rather gloomy introduction, it is clear that ion-selective electrodes have important roles to play in clinical medicine. There are several positive features of ion-selective electrodes that lead to this conclusion:

1. They give a direct index of activity rather than concentration.
2. They can be used to make very rapid determinations on small whole-blood samples.
3. Developments may make possible determination of important species which cannot be done by other methods.

Exploitation of one or more of these advantages, where required in clinical care, rather than an attempt to compete on all fronts, might lead to a more rational application of ion-selective electrodes.

II. CLINICAL REQUIREMENTS

A. Activity or Concentration Measurements?

The inorganic plasma constituents can exist in the following forms:

1. Ionized
2. Undissociated ion pairs
3. Bound to organic anions
4. Bound to plasma proteins

An ion-selective electrode responds to changes in the activity of the ionized fraction. Properly calibrated with carefully designed standards, the electrode will give a measure of the activity of a given ionic species. The differences between its readings and those of the flame photometer, which are expressed as concentrations in the total plasma volume, are then, of course, real differences and not errors.

The old problem of whether biologists are interested in concentration or activity must now be considered. The comment expressed by Mohan and Bates[7] that "concentration is the medically significant quantity" is an oversimplication, which, we feel, derives more from the limitations of existing techniques than from biological significance. In reality, both measurements can be of interest, depending upon the circumstances. The problem is in many ways analogous to that of the pH as opposed to "total titratable acidity" or P_{O_2} as opposed to "available oxygen." When a clinician is trying to estimate the total body reserves of an element or to assess the extent of depletion by urinary loss, measurements of total concentration are clearly required. On the other hand, the biological membranes of the brain, the nervous system, the heart muscle, and the glandular cells act in a similar way to ion-selective electrodes, setting up membrane potentials in relation to the ionic activities on either side of those membranes. These potentials affect the function of the cells. A change in extracellular potassium affects both the heart rhythm and its strength of contraction, and the potassium activity is the important quantity here. Levels of intracellular calcium ions as low as 5×10^{-3} M are extremely important in triggering the contraction of heart muscle. Whether or not activity has any real meaning when considering these very low levels in a system with the fine structure of heart muscle is a separate problem, but the mechanism clearly involves the liberation of small numbers of active calcium ions from structures within the cell.

In many instances, therefore, it is the activity that is the most important in biology,[5,6] since this is what the cells "see" as their ionic environment.

For the simple monovalent cations, binding and complexing are generally considered to be negligible.[8] Experiments using venous occlusion to bring about ultrafiltration in vivo[9] were reported to show protein binding of 10% for sodium and 20% for potassium, but these claims were subsequently modified.[10] Studies with ion-selective glass electrodes by Moore and Wilson[5] and Dahms et al.[6] indicated that binding was not significant, and Ebden[11] found generally no evidence of ion binding for potassium but reported occasional anomalous low activities in some plasma samples. Our own study of potassium measurement (described later) with ion-selective electrodes confirms this view.

For the divalent cations, on the other hand, protein binding and complexing with organic ions are significant. In 1934, a frog's heart was used as a bioassay system[12] for calcium. This showed that calcium in the plasma was less effective than an equivalent concentration in aqueous solution, thus providing the first evidence of protein binding for calcium. More than 50% of the serum calcium is complexed, the majority of it to protein. The total calcium, therefore, changes as the concentration of plasma protein varies in disease[13] or with posture, due to redistribution of body water[14,15] or with prolonged venous occlusion.[16] Various methods of calculating the ionized fraction have been suggested by allowing for the measured level of serum protein,[17] but the extent of protein binding also depends very critically on the pH of the plasma at that time and in that part of the circulation. These attempts at deriving ionized calcium are, at best, approximations and can be quite unhelpful when, for instance, the ionized calcium is reduced by complexation with citrate following blood transfusion,[18] with a marked reduction in myocardial contractility. Magnesium is also partly protein bound, but, at present, there is neither the requirement nor the technology to measure this ion with an ion-selective electrode.

If we now consider the activity of the ionized fraction, there is again a marked difference between the monovalent cations, sodium and potassium, and the divalent calcium ion. Not only are sodium and potassium free in solution, they are also fully dissociated and appear to have activity coefficients in plasma essentially the same as for aqueous solutions of the same ionic strength.[7] Here the homeostatic mechanisms of the body help. The osmolality* of the plasma is fairly well controlled. The ions are in solution against a fairly constant background of an ionic strength of close to 0.15 *M*, and in this region of concentration, small changes in the ionic strength do not produce marked changes in the activity.

Activity coefficients for calcium are a much more difficult problem. Although it has been said that activity coefficients of calcium chloride in mixed solutions are the same as those which would hold for pure calcium chloride solutions,[19] we are not aware of any direct studies in blood. In addition to this, the variability in the protein-bound and complexed fractions and the problem of incomplete dissociation make the relationship between calcium activity and concentration extremely unpredictable.

Clearly, then, in the case of calcium, a successful electrode would be of importance in its own right. It would produce information of great biological and clinical significance which is not available in any other way. In contrast, the relationship between activity and concentration for sodium and potassium is simple and to a large extent predictable. The ion-selective electrode provides little extra information to the clinician, and its use cannot really be justified on the theoretical grounds that it is an activity measurement.

B. Speed and Convenience

The other advantage that the ion-selective electrode has to offer is the convenience

* Total electrolyte molality (Ed.).

of a rapid result from a direct reading on whole blood. The value of an activity measurement of calcium is of such overwhelming significance that, as has been pointed out, a successful ion-selective electrode would justify its existence on these grounds alone. Speed and convenience would be a bonus.

Although there is no fundamental requirement for activity measurements of the monovalent cations, there are clinical situations in which rapid and frequent potassium measurements are required. Routine clinical laboratories provide multiple measurements on one sample of blood. A typical grouping would be sodium, potassium, and urea, together with a variable number of others chosen from chloride, sugar, and creatinine. The selected group of tests might be carried out once, as a screening test on first seeing the patient, and at intervals on outpatients with, for example, chronic renal and cardiac disease. For inpatients, the same tests may be performed daily in postoperative or acute medical cases and sometimes more frequently; however, with a few exceptions, there is no need to to measure sodium, chloride, urea, and creatinine more often than once a day. Potassium, on the other hand, can be very variable and its level critical in a minority of patients, for example, severe diabetics or those in intensive care unit following cardiac surgery or coronary occlusion.

We have already expressed the opinion that ion-selective electrodes can offer little advantage over the conventional flame photometer for long, routine runs of measurements. However, we feel that ion-selective electrodes are the instruments of choice for the measurements that are required rapidly and not in combination with a great range of other measurements.

Potassium is a good example of this type of requirement. Ideally, diagnosis of dangerous levels and frequent monitoring to facilitate rapid but safe correction requires on-the-spot measurement, not in combination with other electrolytes but more often at the same time as blood gas and acid-base measurements. In addition, the normal range of potassium (Table 1) represents nearly 12 mV change in the potential of an ion-selective electrode.

For these reasons, we consider there is a requirement for a potassium electrode and that this requirement can be fulfilled. The practical details of a potassium electrode for the intensive care unit are described in a later section.

Sodium measurements with glass electrodes are difficult in practice and cannot be justified under any of the requirements discussed. Consequently, we feel that ion-selective electrodes have little to offer in sodium estimations for clinical use.

III. PRACTICAL CONSIDERATIONS

Many of the highly selective membranes available will operate in the clinical range required. Although it would seem a simple matter for the average laboratory worker to adapt these for biological measurement, some points are worth noting.

A. Accuracy and Reproducibility

The normal levels of the various plasma electrolytes are shown in Table 1. The total normal range for sodium covers little more than 10 mmol/ℓ and for potassium about 2 mmol/ℓ. For most clinical purposes, a tolerance of 2 mmol/ℓ for sodium and 0.2 mmol/ℓ for potassium would be realistic. Reproducibility and discrimination are often more important than absolute accuracy in either concentration or activity terms — the detection of a trend by examining the results from two samples at a suitable time interval is often useful.

B. Expression of Results

Current medical practice uses millimoles per liter as a means of expressing concentrations. Despite some theoretical advantages, pIon scales would be of little value to the clinician. Expressing electrode results in activity terms is unnecessary and could cause dangerous confusion.

The majority of clinicians are well aware that although they express measurements in concentrations, they are dealing with a form of activity scale. The results are not usually for equilibrium calculations, and so the actual activity coefficient is immaterial. One exception is found in the use of the Henderson-Hasselbalch equation for the calculations concerning the bicarbonate ion in the blood plasma. This takes the familiar form,

$$pH = pK_1' + \log \frac{(HCO_3^-)}{(\alpha P_{CO_2})}$$

where c = calcium ion concentration.
Here there can be conflict in the scales used. pH is a conventional activity, and P_{CO_2} is an index of fugacity. When the bicarbonate is required in concentration terms, an activity correction must be applied. This is usually incorporated in the value for the first dissociation constant (approximately 6.1). While this certainly gives the "right" answer in simple salt solutions at an ionic strength of 0.15 *M* it does not necessarily give a reliable index of, for instance, the total CO_2 in the plasma. The discrepancies have been examined by Linden and Norman.[20]

This equation has been the cause of surprisingly bitter arguments among physiologists and clinicians ever since blood gas electrodes came into general use. The majority of the arguments arise from a failure to define terms at the outset and, specifically, from the confusion of scales. Similar problems will clearly arise with the more widespread use of ion-selective electrodes, particularly where concentrations are also available. We feel that considerable trouble could be avoided if the firms concerned with ion-selective electrodes concentrated on the scientific basis of the measurement techniques that they offer rather than on promotional literature that is no more informative than the glossy drug advertisements.

The simplest convention for the monovalent cations would be to use mixed standards at an ionic strength of 0.15 *M* and to report the results as concentration in relation to these (activity could be readily calculated, if required).

C. Liquid-junction Potentials

As with pH measurements, the liquid-junction potentials formed between the reference bridge system and biological fluids introduce uncertainty into the measurements. The narrow range for many biological measurements does not help the situation. Taking the clinical ranges from Table 1, the change in voltage (assuming a theoretical electrode) of the cell for sodium would be 2.40 mV; for potassium, 11.35 mV, and for calcium, 2.01 mV from one extreme to the other of normal. The tolerances that we have suggested for an individual measurement mean that the sodium cell must be reproducible to better than 0.4 mV and potassium to better than 1.3 mV. (These ranges and tolerances may vary slightly between different laboratories and clinicians, but the figures illustrate the problem.)

Clearly, the small changes in EMF which are involved necessitate the choice of a highly reproducible liquid junction. It is not quite so apparent that the actual magnitude of the E_j is also important. If the E_j were the same whether the bridge solution was in contact with a blood sample or with a standard solution, then the absolute magnitude would be immaterial, since the effect of the E_j would be absorbed in calibration, along with the other potentials of the cell. However, any change in E_j that occurs when a standard solution is replaced by a test solution (the so-called "residual

E_j'') has an important effect on the results. On a logarithmic plot scale, the discrepancy would be a constant error in one direction or the other, depending on the sign of the residual E_j. However, when the results are plotted on linear scales as concentration relative to the standards or as activity against actual measured concentration in the test solution, the residual E_j affects the slope of the line. Thus, the residual E_j is incorporated in the activity coefficient for the ion in the test solution, as determined by simultaneous electrode and conventional concentration measurements.

The various forms of constrained diffusion junction do not give such consistent results with biological solutions (e.g., blood) as they do with simple salt solutions. Red blood cells are in suspension and settle out slowly during the measurement. In contact with hypertonic bridge solutions, the cells crenate and may form plugs that fall through the junction. Highly reproducible junctions for simple solutions may be formed in narrow parallel tubes, preferably with the heavier concentrated bridge solution underneath the test solution. This classical free-diffusion junction is usually used for high-quality pH determinations. It does not appear to be quite so reproducible with blood. In practice, the junction of a capillary filled with blood dipping vertically down into a chamber filled with the bridge solution is one of the most satisfactory. This is widely used in many blood pH systems, e.g., Radiometer K 497, and was the one used in our own potassium cell.[22] Other junctions that have been utilized are a cellophane membrane separating KCl from the test solution (Instrumentation Laboratory Ltd., Altrimcham, Cheshire, U.K. Automated Blood Gas Analyzer) and a continuous flowing junction (Orion Research Calcium, Cambridge, Mass.). This last was used originally by Lamb and Larson[21] and has some theoretical advantages. It is difficult to achieve in simple electrode systems.

D. Temperature Control

With the noncomplexed ions, there is no fundamental reason for choosing any particular temperature for the measurement. In order to achieve the level of reproducibility required, however, it is extremely doubtful whether a simple room temperature system would suffice. Blood samples taken from the patient may be either 37°C (body temperature) or, particularly if the same sample is to be used for blood gas analysis, may have been cooled in ice water. It is often convenient to thermostat the electrode at 37°C. In the case of calcium measurement, where the activity may be significantly influenced by temperature change, measurement at 37°C is mandatory.

E. Cuvette Design

The design of cuvettes for ion-selective electrodes operating in biological fluids is more critical than for blood gas electrodes. The problem appears to stem from the absence of buffering for the ions in the calibrating and test solutions. The wash-out of all parts of the cuvette and sensing membrane therefore becomes extremely important. The volume of blood available may be small, particularly from children and newborn babies, and the smaller the cuvette volume, the better. The actual volume of the cuvette does not necessarily give an indication of the sample size that is necessary to achieve a complete wash-out. The cuvette in our potassium electrode[22] has a volume of 50μl and gives adequate wash-out with samples of 1 ml. Figure 1 illustrates a design fault in a cuvette, which necessitates washing through with several milliliters of standards or test solution. This problem also stresses the thermostatting and causes a longer read-out time. The more closely a cuvette approaches the uniform section of a capillary, the better is its wash-out.

F. Standards and Calibration

The use of standards with a constant activity coefficient has obvious practical advantages. This is achieved by making up standards in a background of constant ionic strength which approximates that of plasma.[7] This has the additional advantage of minimizing the residual liquid-junction potential. The validity of this approach has been confirmed in practice by the finding that ion-selective electrodes show for any given concentration of potassium the same voltage in the calibrating solution and in blood. If an alternative bridge solution were to be used, this simple confirmatory experiment would have to be repeated to exclude a significant residual E_j.

Mixed standards can be designed to check on the development of cross-sensitivity with other ions in the plasma. Potassium standards containing sodium chloride, for example, provide a frequent check on the selectivity of the ion-selective electrode in use.

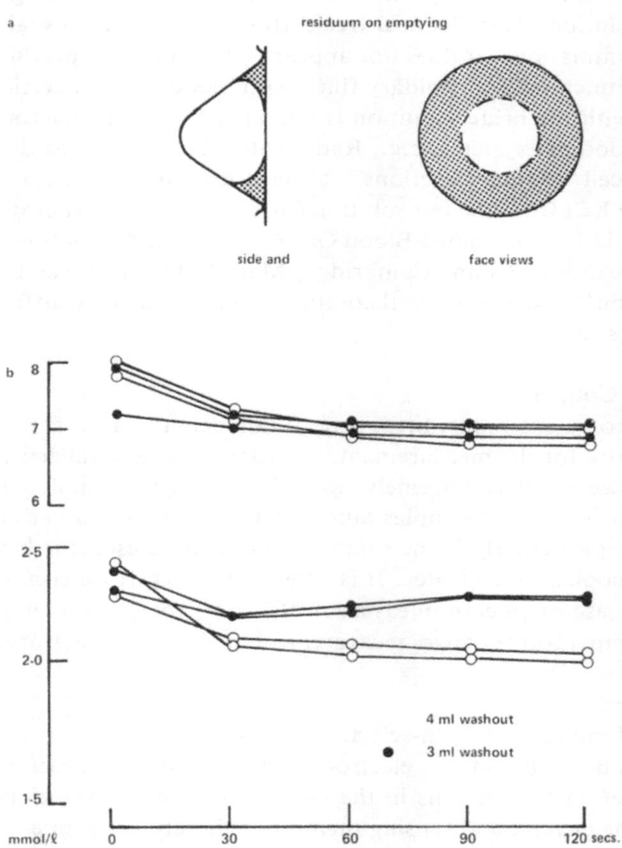

FIGURE 1. (a.) A cuvette of 200 µl with poor wash-out characteristics, as seen with blood or standard containing dye. (b.) The time course of meter readings on changing between 2 and 7 mmol/ℓ KCl activity standards. The first part of the curve (0 to 30 sec) is due to the temperature change after introducing standards at room temperature into a thermostatically controlled cell at 37°C. This is due to a change in E° of the internal Ag/AgCl reference electrode and therefore is of the same direction and magnitude, irrespective of K^+ concentration (about 2 mV). Changes between 30 and 90 sec are due to mixing of high and low standards. This is a concentration-determined change and is therefore seen more clearly in the low standard, due to the logarithmic arrangement of the y-axis. Both these effects are maximal in a cuvette with poor wash-out characteristics.

A further advantage of isotonic standards is that they avoid the risk of error due to hemolysis which might occur as standards and blood samples follow each other in the cuvette.

G. Meter Read-out

Most firms offering ion-selective electrodes also supply suitable meters and measurement systems, as discussed in Chapter 2, Volume I. However, many biologists wishing to investigate the possibilities of ion-selective electrodes may wish to construct their own or adapt existing meters. Many laboratories have blood gas-measuring equipment which is not in routine use, having been superseded by more modern equipment. Any meter with a P_{CO_2} scale, e.g., Electronic Instruments Ltd. (Cheitsey, Surrey, U.K.) 48C or Radiometer (Copenhagen) pHM27 with blood gas analyzer, can be used as a direct reading instrument. For potassium, the conversion factor is simply 10^{-1}. Some relabeling may be convenient for other ions. It is sometimes necessary to provide a small bucking voltage from a mercury battery in the reference connection to get the electrode voltage within the range of back off provided by the meter.

H. Electrical Interference

Electrical interference is a major problem with high-resistance microelectrodes, and effective screening is the only real solution. With lower-resistance miniature electrodes, satisfactory recordings can, however, be made without screening and in the presence of other electrical equipment. We have found it essential to use a balanced, i.e., differential, input Field Effect Transistor (F.E.T.) probe unit physically close to the electrodes. In animal work, the common of this probe is treated as an earth point for the preparation and metal structures in its vicinity.

The regulations concerning patient safety impose constraints on recording systems used in man. Simple, direct indicating meters are conveniently made fully floating from internal batteries. The two electrode inputs can be protected by high resistances and the common connected to the patient. The patient is usually not grounded, as this can be hazardous when other transducers are being used at the same time. Obtaining a recording from a floating amplifier of this kind requires an isolating circuit, since the output is relative to the common line, and it is, therefore, dangerous to connect this to most conventional recorders.

Electrical interference is commonly 50 or 60 Hz mains interference and radio frequency. In some applications, 50 Hz may encroach on the frequency required from the electrode and is better eliminated at source. Both forms of interference can, however, cause DC shifts, which can be overlooked by the unwary. Many electrode cells can act as partial rectifiers. Although the sounds of orchestral music emanating from the vibrating pen of the recorder can be entertaining, the phenomenon may not always be recognized, particularly when testing electrodes on a direct recording meter. Strong radio-frequency signals in the immediate vicinity, e.g., radio call systems, are the most troublesome. Very strong interference may start to swamp the input of the amplifier. Some commercial solid state meters are very prone to this. An electrode which will read satisfactorily on a vibrating capacitor electrometer or on a dual FET with good common mode rejection may show a wandering voltage and poor scale length on a solid state commercial meter. This is traceable to hum acting at the input, i.e., the electrode being inadequately screened for this type of meter.

Our own approach is to rely on common mode rejection at the input with no attempt at filtering, i.e., to keep the input capacitance as low as possible. The raw signal from the amplifier is then monitored to check on the interference level before being suitably smoothed and recorded. In this way, a continuous recording up to 20 Hz can be obtained with a noise level of ∿30 μV peak to peak (under ordinary laboratory conditions) with an electrode resistance of around 10 MΩ.

I. Chemical Interference

Interfering ions are considered elsewhere in this study (Chapter 1, Volume I). We are concerned in this chapter with problems peculiar to biological solutions. Measurements in bile, urine, and the contents of the gut present special problems. However, the performance of a polyvinyl chloride (PVC)-based electrode is not adversely affected by plasma constituents. The clinical range of bilirubin, cholesterol, and urea did not affect the electrode results in our potassium series. We have also found that anesthetic levels of the barbiturates do not affect the electrode nor the blood gases, including carbon dioxide. Many of the gaseous anesthetics, however, act as powerful solvents for PVC - based membranes. We have seen no reports of any effects of these agents at anesthetic concentrations, but there could conceivably be a danger of their causing deterioration of the membrane. Orion (Cambridge, Mass.) has given warnings that contaminating samples with oil can affect their calcium electrode.

IV. ION-SELECTIVE ELECTRODE FOR MEASUREMENT OF PLASMA POTASSIUM

A potassium electrode was developed originally with the object of determining activities in biological fluids.[22] A combination of valinomycin and tetraphenylborate[23] is used in a PVC membrane to combine high selectivity with improved conductivity and anion rejection. The composition of the membrane is given in Table 2. We use PVC tubing of 4 mm outside diameter as the basis of the ion-selective half cell and dip cast a PVC membrane onto a porous ceramic plug. A silver/silver chloride internal reference electrode and 4 m*M/ℓ* internal reference solution complete the half cell.

The membrane is surrounded by a PVC washer, glued on so that the junction is electrically tight (Figure 2). This is held in a thermostatted perspex holder (Figure 3) so that the washer and membrane form one side of a cuvette. It is important that the whole of the membrane is exposed to solution so that no trapping of the membrane edge can occur. Other features of the cuvette design have already been discussed. The circuit is completed by dipping the capillary into the saturated KCl bridge of a Radiometer (Copenhagen) K497 calomel reference electrode. The reason for this particular reference electrode, the configuration of the liquid junction, and the meter read-out used has already been discussed in Section III. G.

TABLE 2

The Composition of the Potassium-selective Membrane

Component	Amount
Valinomycin	0.0015 g
Potassium Tetraphenyl Borate	0.000025 g
Dioctyladipate	0.15 g
(Bis-2-ethylhexyladipate)	Variable
Nitrobenzene	0.05 g
High-molecular-weight PVC	0.075 g
Tetrahydrofuran	3 ml

Adapted from Band, D. M. and Kratochvil, J., *J. Physiol.* (London), 265, 5P, 1977.

Standards of potassium chloride of 1 to 10 m*M/ℓ* made up to ionic strength of 0.15 mol/ℓ with sodium chloride were used (Table 3) to study the electrode response in the clinical range, and Figure 4 shows a calibration repeated after an interval of 3 days. For routine use, the electrode is calibrated with a low standard, e.g., 2 mmol/ℓ KCl and the span adjusted with 8 mmol/ℓ KCl. Most electrodes show an initial response of 97 to 99% of the theoretical Nernst relationship. This is maintained for at least 18 months if the electrode is stored dry, and a useful life of 3 years has been reported.[24] With use, however, enough of the ion exchanger is leached from these membranes over a few months to diminish the response. The use of mixed standards ensures that a deterioration in selectivity will not escape detection.

The result of the first 50 clinical blood samples is shown in Figure 5. Both ion-selective electrode measurements on whole blood and flame photometer measurements on separated plasma are the result of duplicate readings, and the reproducibility was of the order of 0.02 mmol/ℓ.

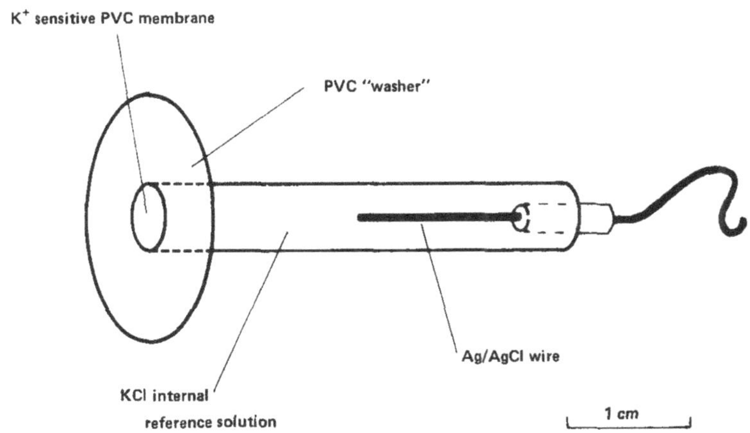

FIGURE 2. The ion-selective half cell for use in the perspex cell (Figure 3).

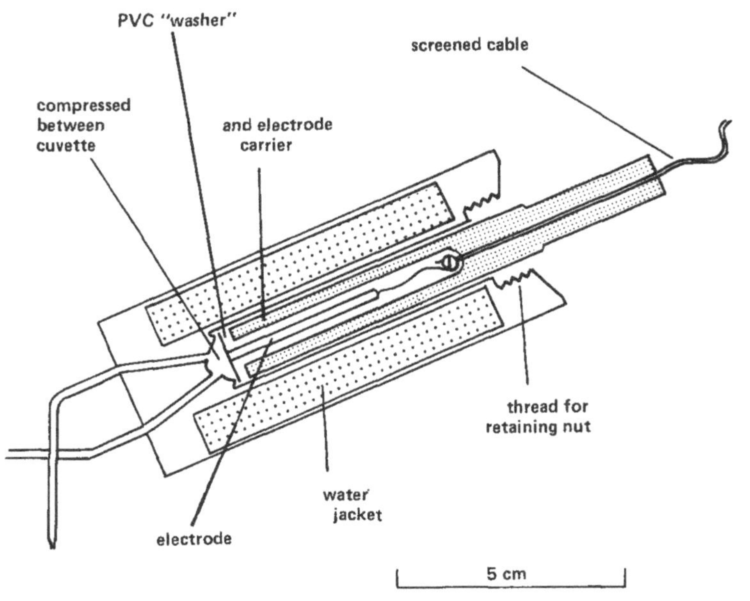

FIGURE 3. Thermostatted perspex holder containing the ion-selective half cell.

TABLE 3

Composition of Constant Activity Potassium Standards Used in Experimental Studies of Ion-selective Electrodes

KCl (mmol/ℓ)	NaCl (mmol/ℓ)	Ionic Strength (mmol/ℓ)
1	149	150
1.5	148.5	150
2	148	150
2.5	147.5	150
3	147	150
4	146	150
5	145	150
7	143	150
10	140	150

Note: For routine calibration, a low and high standard are sufficient, the low (2.0 or 2.5 mmol/l) being chosen to suit the zero point of the meter and a high standard to adjust the span.

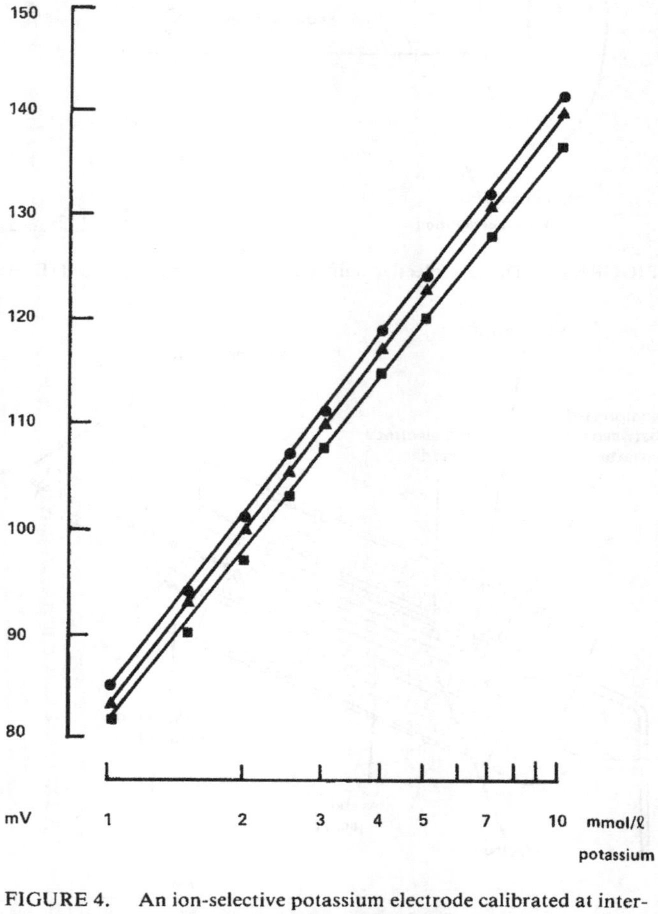

FIGURE 4. An ion-selective potassium electrode calibrated at intervals of 3 days with mixed KCl/NaCl standards (Table 3).

There is a systematic discrepancy, with the ion-selective electrode reading being 0.2 mmol/l higher than the flame photometer. This is probably because of the space-occupying effect of the plasma proteins (4% by volume of the sample), reducing the apparent concentration of the flame photometer measurements.[5,6] We have tested this hypothesis by comparing the magnitude of the discrepancy with the plasma protein concentration in each of 54 samples and have found a positive correlation. Any protein binding would tend to shift the line in the other direction and flatten the slope. Our experimental findings confirm the view that potassium is in free solution in plasma.

After allowing for the discrepancy due to the volume effect of the plasma proteins, the results of the electrode and concentration measurements were very close to a line of identity when the electrode was calibrated using concentration standards of similar ionic strength to the blood plasma. We infer from this that the free diffusion, saturated-KCl liquid junction chosen produced only a very small residual E_j on changing from the isotonic standards to the whole blood. On the other hand, the actual magnitude of the E_j formed between saturated KCl and the whole blood has for many years been a source of considerable theoretical argument (in relation mainly to blood pH measurements), and the true activity coefficient for K^+ in plasma is, of course, unknown. It could, therefore, be argued that the line of identity was merely a coincidence, the residual E_j effect canceling the difference in the activity coefficients for our standard solutions and the blood. However, the choice of liquid junction has already been discussed, and it would seem more reasonable to accept these results as evidence that any residual E_j in this junction is minimal and that the activity coefficient for potassium in whole blood is essentially the same as in 0.15 mol/l NaCl. It is worth noting that a residual E_j of 3 mV would alter the slope of the line from unity to 1.25, thus introducing an error of 1 mmol/ℓ in the potassium result at the upper end of the range. In our view, before introducing any junction configuration or change in bridge solution for measurements in blood or plasma, this sort of direct comparison must be repeated.

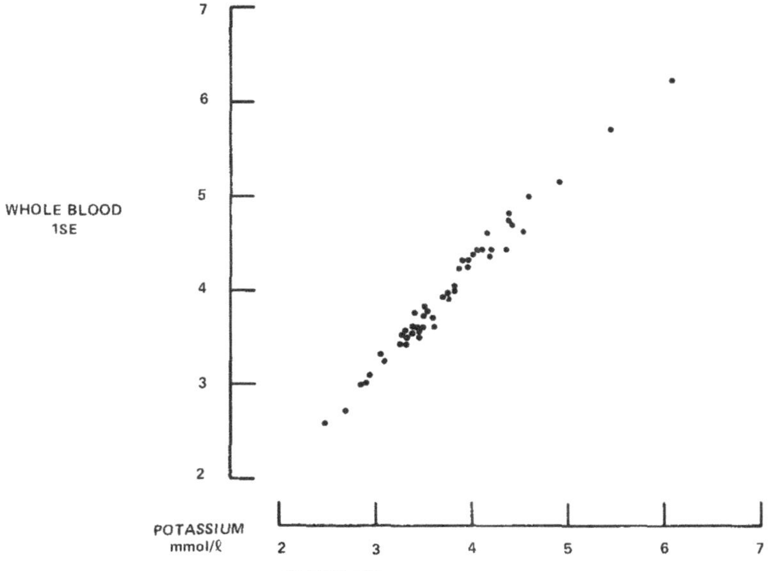

FIGURE 5. Fifty clinical blood samples. Potassium measured by an ion-selective electrode and a flame photometer.

A comparison of ion-selective electrode measurements on whole blood and separated plasma from the same sample was also made (Figure 6), which confirms that the electrode measures potassium in either specimen equally well and that there is no significant suspension effect due to the red cell at the liquid junction. A further series of 115 blood samples were studied, taken from patients in the intensive care unit (Figure 7). The same arguments apply, but in this study, the scatter is rather greater. Some of the results where the ion-selective electrode reads high compared with the flame photome-

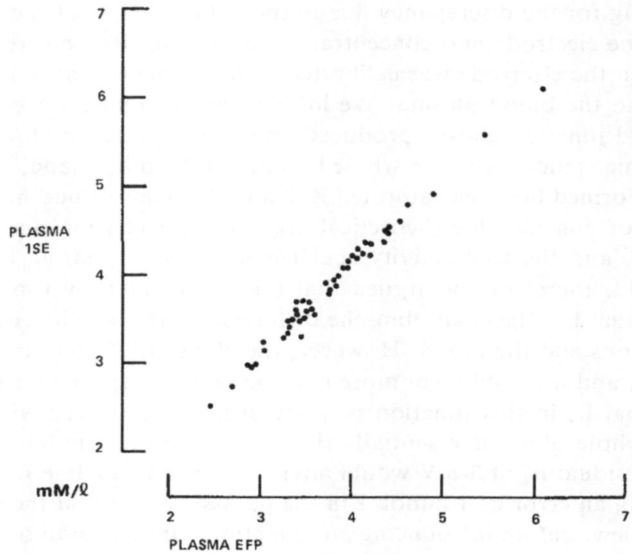

FIGURE 6. Fifty clinical blood samples. Ion-selective electrode measurement on plasma and whole blood are compared.

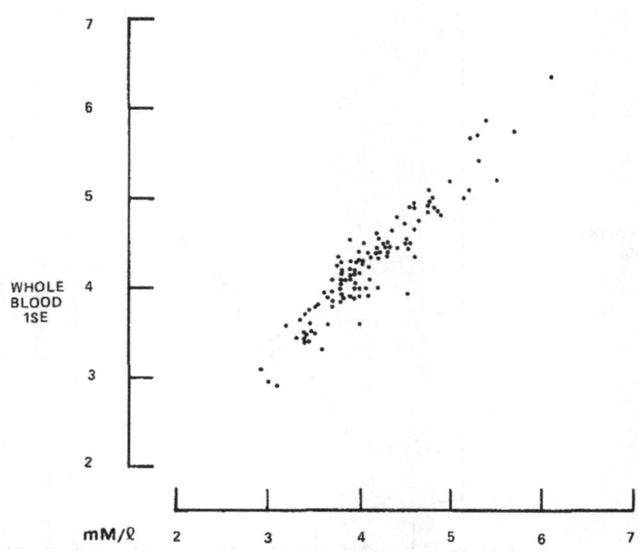

FIGURE 7. One hundred and fifteen blood samples from the intensive care unit.

ter are due to an excessive content of space-occupying molecules. One, for example, was from a diabetic girl whose serum was milky due to the high content of lipids. In this case, of course, it is the electrode which produces the more useful clinical information, since it detects activity in plasma water, the quantity of which affects the cells. All these blood samples were investigated in the intensive therapy unit. The physicians there rapidly realized the possibilities of the method and took over the electrode cell for their own use. Its value to them lies in the fact that it is always ready to give an immediate answer when a question arises, particularly in the middle of the night or when a patient is in the operating theater. For this particular application, the maintenance of a flame photometer for the 20 or so emergency samples that may be required in the course of 24 hr would probably not be justified. The necessity for centrifuging the samples and the warm-up time of the machine also weigh against the flame photometer.

The Orion SS-30 sodium/potassium analyzer, according to its specification, fulfills some of the criteria we have outlined. However, the inclusion of the sodium electrode and the degree of automation and digital read-out may have been invaluable in space and within the budget of the National Aeronautics and Space Administration (for whom Orion originally developed the Space-Stat System) but are not necessarily required by clinicians with their feet on the ground. In practice, automatic and semiautomatic instruments are, of course, quick to operate when runs of measurements are to be made, and malfunctions can be detected before any dubious results are used. However, a clinician making just one or two spot readings who must be absolutely certain of his measurements is usually more satisfied with an instrument that directly he calibrates himself and that allows him to see what is happening.

V. CALCIUM ELECTRODES

Calcium electrodes have been available for more than 10 years, and there has been a proliferating literature on their use. Some of the early electrodes were not entirely satisfactory. In the literature provided with the recent model Space-Stat 20 ionized calcium analyzer, Orion Bio-Medical (Cambridge, Mass.) stated of its earlier electrodes that the manual method of electrode analysis required a highly skilled technician, so their use was mainly confined to research groups and teaching hospitals. The newer Space-Stat 20 uses an automatic flow-through system. The reference electrode is a silver chloride pellet through which a potassium chloride reference solution (saturated with silver ions) is pumped. A relative accuracy of ± 0.02 meq/ℓ (i.e., approximately 1% of usual reading of 2 meq/ℓ) is claimed. (This would presumably be equivalent to a pH cell giving results ± 0.002 pH). Such precision is required, since the normal range of expected ionized calcium would be represented by only 3 mV change in the electrode voltage (equivalent to 0.05 pH units).

Orion Research has clearly aimed the Space Stat system at the purely clinical market. In so doing, it has simplified the problem of calcium measurement. All results are expressed in concentration terms, and even the Nernst equation quoted in their literature is given as

$$E = E_0 + s \log C \quad (1)$$

The apparatus contains an internal standard of 1.00 mmol/l calcium ion (actually described as 2.00 meq/ℓ) with a background of sodium chloride and a Tris hydroxymethyl) aminomethane buffer to simulate the ionic composition and pH of blood.

If the system lives up to its specification, it will provide information that can be interpreted in the light of clinical experience and should prove extremely valuable.

Instruments of this type do not, however, attack the fundamental problem. They cannot give results of "ionized calcium concentration," as is claimed, but, in reality, give numbers on a scale that is operationally defined by the measuring system used. This is not to decry the usefulness of the approach, since any index of calcium activity is better than total calcium measurements, which may be completely meaningless in the clinical context.

At the heart of the problem of calcium measurements is the problem of calcium standards. First, calcium activity standards in mixed electrolytes are required before the cross-sensitivity of electrodes can be determined. Only then can these standards be used for establishing some scale of activity in the biological fluids. The simplification possible with the fully dissociated monovalent cations is not permissible — in a sense, the problem with calcium is much closer to that of pH measurement.

Some progress has been made; calcium standards have been offered by Covington and Robinson[25] and by Mohan and Bates.[7] However, further work is required. The evaluation of the sensitivity of any calcium-selective membrane at present relies on the investigation of the slope of the electrode's response to variations in calcium against a constant background of other ions, the aim being to hold the activity coefficient constant. Interference by other components of the biological fluid is equally difficult to evalute, and here there is considerable danger of falling into a circular argument — to distinguish between "poisoning" of the electrode and depression of the activity of the ion is not easy.

The problem of residual liquid-junction potential again emerges. With the very small voltage range represented by the variation in calcium, this becomes even more critical in defining the activity scale. Our own experience with calcium electrodes for medical purposes has been confined to systems based upon a PVC matrix.

The success of the PVC-potassium electrodes in blood led to attempts to incorporate various ligands into PVC for calcium measurements. The cell used was identical with that for potassium measurements (Figure 3). The calcium ionophore A23187 (Ely Lilly, Indianapolis, Ind.) gave electrodes that showed an initial response to calcium but deteriorated over a few days of use. A polymer incorporating Ca-dioctylphenylphosphate in dioctylphenylphosphonate proved more successful. A calibration curve is shown in Figure 8 for this membrane, and the range of standards is given in Table 4. The effect of magnesium on this membrane is peculiar. In magnesium standards (without calcium), the electrode voltage is extremely unstable, and on retesting in calcium standards, its response is then impaired for several minutes. In mixed calcium magnesium standards (calcium 0.1 mmol to 1 mmol/ℓ, magnesium 1 mmol/ℓ), Fleet et al.[26] found that the magnesium affected the time taken to reach equilibrium voltage rather than having a direct effect on the EMF.

In summary, the calcium electrode is not in widespread clinical use nor do we feel it is at a stage where it can be put into routine service. The measurement has not previously been available, and, therefore, the clinician has no real guidance as to the significance of the results reported to him. The technology is now at a stage, however, where careful critical evaluation first of what governs the state of calcium in the blood and then the application of this knowledge to clinical care would be of great interest.

VI. CONTINUOUS IN VIVO MEASUREMENT

Because ion-selective electrodes can produce an "immediate" measurement on a sample in whole blood without consumption of the sample, it is possible to adapt them to give continuous measurement of the particular ion species in the flowing blood. This was first done for sodium[27] and potassium,[28] using glass electrodes. The value of this work was limited by problems of high-resistance glass electrodes and with streaming potentials producing erratic results. Continuous pH measurements were made by Band and Semple[29] in the cat and in man.[30]

Today, 10 years later, the continuous in vivo monitoring of specific ions in patients is still experimental. Only where the in vitro techniques are well evaluated and the measurements of sufficient clinical value would such an invasive technique be justified. Above all, the frequency with which information is required must be sufficient to merit the problems encountered.

pH measurement in vivo has continued to be used but has never become routine. Staehelin et al.[31] used a glass electrode on a cardiac catheter tip and a skin electrode as the reference. They had a high failure rate in animal experiments (12 out of 32

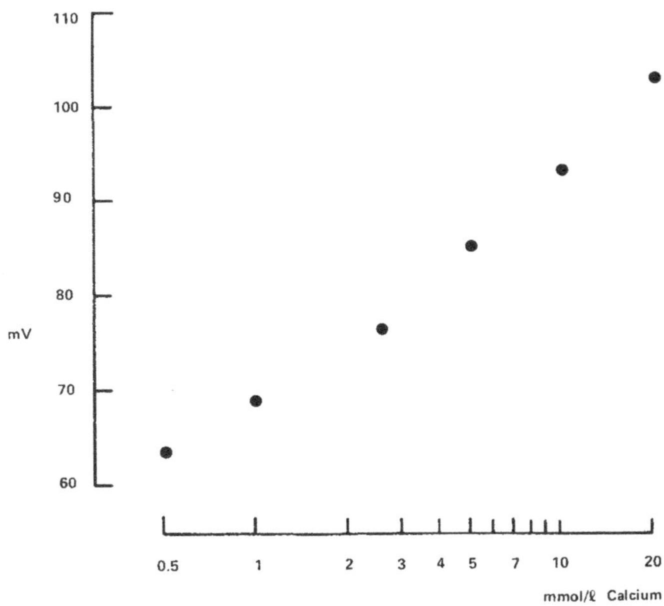

FIGURE 8. Calcium electrode calibrated with $CaCl_2/NaCl$ standards (Table 4).

TABLE 4

Composition of Experimental Standards for Use with a Calcium Electrode

Calcium[a]	Magnesium[a]	Potassium[a]	Sodium[a]
0.5	0	0	140
1.0	0	0	140
2.5	0	0	140
5.0	0	0	140
10.0	0	0	140
20.0	0	0	140
0	1.0	0	140
0	5.0	0	140
0	10.0	0	140
0.5	1.0	3.0	140
5.0	1.0	3.0	140

[a] Concentrations in millimoles per liter.

dogs), due to technical failures of various sorts. Filler[32,33] has reported his experience with continuous muscle pH monitoring in infants, and, more recently, Le Blanc et al.[34] have used a polymer membrane pH electrode on a catheter tip. We[35] have used K^+-PVC membranes mounted on catheter tips to follow potassium changes over many hours in animals and in patients.

The catheters are 1.65 mm in diameter, and the basic construction and electrochemical configurations are illustrated in Figure 9. Continuous potassium measurements can be obtained with a rapidity of response and discrimination (Figure 10) more than adequate for biological use. The calibration in vivo shows a theoretical relationship between the change in potential and the logarithm of the potassium concentration (Figure 11). These catheters will follow changes in the blood without clot or protein deposition on the membrane or loss of response (Figure 12). We occasionally flush the salt bridge in the reference lumen. We have used these devices for physiological experiments on laboratory animals, and we have also used them in patients on a few occasions in circumstances where the potassium may be particularly variable.

This sort of intravascular continuous monitoring has only limited use in specialized units, but if techniques became sufficiently reliable, they could be extremely valuable in cardiac surgery and intensive care units. pH and potassium changes are of great clinical importance in these patients. pO_2 measurements involving a Clarke polarographic electrode mounted on a catheter tip have also been made with rather more overall success,[36] but further discussion is not merited in this study, as this device falls outside the field of ion-selective electrodes.

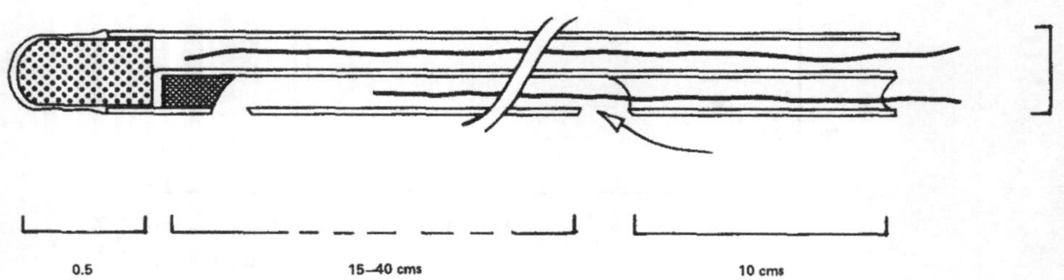

FIGURE 9. Potassium-sensing catheter for in vivo use. These catheters are 1.65 mm in diameter and about 40 cm long.

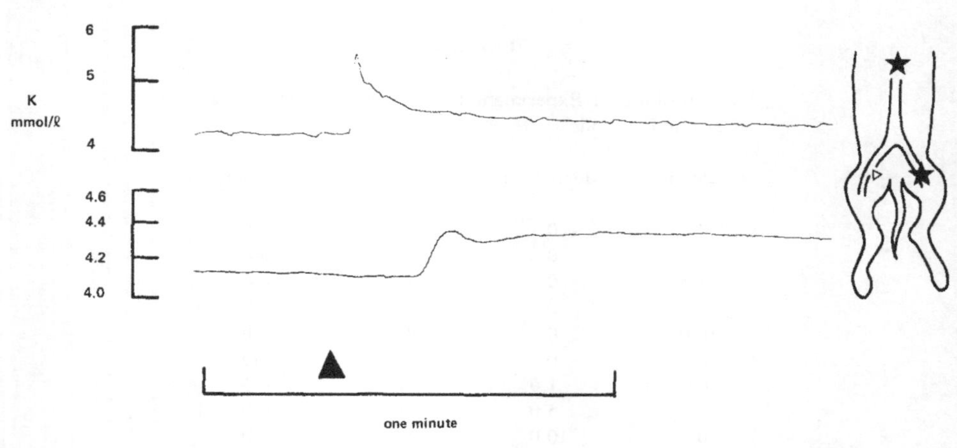

FIGURE 10. Two potassium-selective catheters in the venous circulation of an anesthetized dog. The sensing tips are indicated by an asterisk*. An injection is made of KCl at →, and a response can be seen to the potassium in the upper trace (inferior vena cava). After two capillary circulations, a rise in potassium is seen in blood returning in the femoral vein.

The practical considerations involved in obtaining measurements of this type begin with the need for in vitro evaluation of the ion sensor and its calibration. The second practical problem is calibration in use. In attempts at pH and potassium measurements, this has been achieved by periodic withdrawal of samples (Figure 12). In a membrane with a variable asymmetry potential such as the pH glass electrode, this is difficult to avoid. If there is no asymmetry potential, the stability depends on the reference electrode systems. This is the case with PVC-potassium membranes where, theoretically at least, readings of the reference electrodes against each other should provide all the information needed for accurate calibration.

The means of making the sensing membrane of lower resistance by increasing its surface area without increasing the diameter of an intravascular device has brought the source impedance to manageable levels. This has been achieved by Le Blanc et al.[34] by casting a pH-sensing polymer membrane on silver wire and potassium-sensing PVC membranes onto porous ceramic.[35] Studies of this kind still depend on home-made,

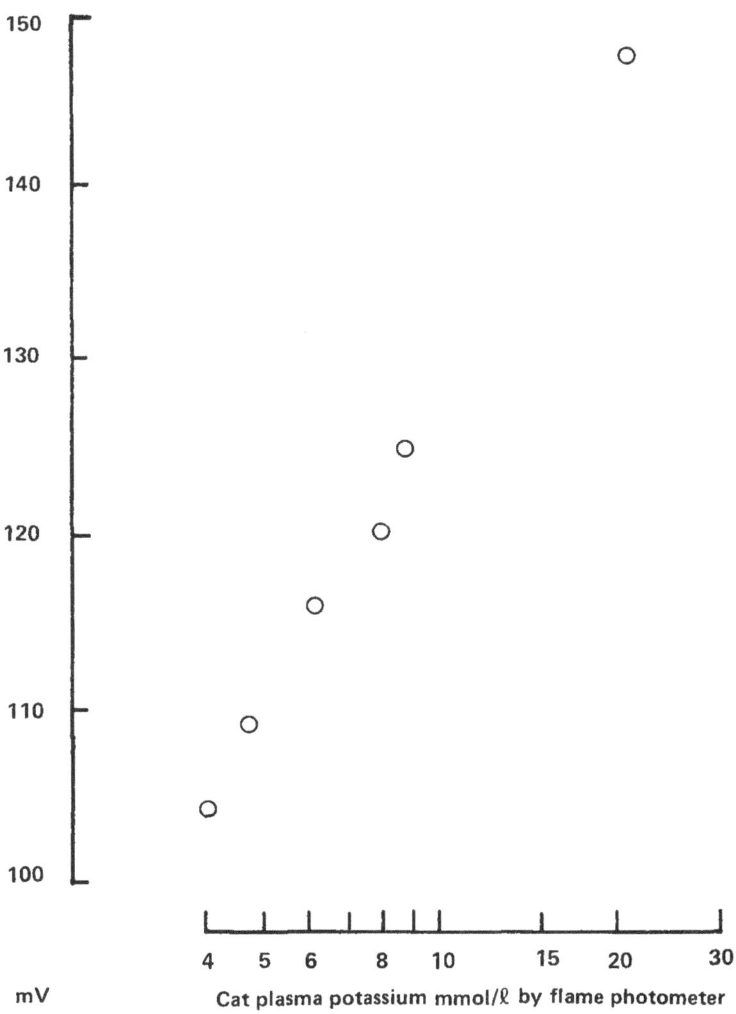

FIGURE 11. A potassium electrode calibrated in vivo. Voltage readings from the catheter are plotted against the simultaneous concentrations in blood samples determined by flame photometer. The plasma potassium was raised progressively by repeated injections of potassium chloride.

purpose-built electrodes and application by trial and error of general principles of electrode measurement.

The basic principles governing the choice of reference system for in vivo measurement are simple. The reference should be as close as possible to the sensing electrode. There should be no intervening structure between the two; e.g., a cell membrane may frequently act as a potassium-selective membrane, and substantial potentials may exist across it. Leakage of bridge solution should not affect the properties of the system being investigated.

A number of additional errors may be introduced into in vivo recordings that are not met in the more controlled conditions in vitro. Picking up spurious bioelectric potentials can be avoided by keeping the reference junction as close as possible to the sensing electrode. Streaming potentials can be a major artifact when recordings are made from the cardiovascular system. The pH electrode used by Band and Semple[29] was blown as a blind-ended glass capillary. This was housed in a section of stainless steel needle tubing which extended as far as the reference electrode. The needle tubing acted as a third electrode and reduced the streaming potential problem to the point where only insignificant voltage changes were seen on complete flow reversal. (The use of a screen in this way but on a much larger scale is sometimes used for industrial electrode combinations.) The walls of blood vessels are electrically conducting, and, for this reason, it is often easier to make satisfactory recordings in vivo than in any in vitro mock-up.

Glass electrodes appear to be free from pressure artifacts. Membrane electrodes, on the other hand, generate large potentials when subjected to pulsatile pressures. It is essential to "splint" these membranes in some way, e.g., by casting directly onto some ceramic.[35] Pulsating pressures affect most liquid junctions to some extent. Some form of constrained diffussion type with a pressurized bridge solution can be used.

VII. MINIATURE AND MICROELECTRODES FOR BIOLOGICAL USE

Electrodes have been used in a variety of biological preparations. The possibility of making measurements of pH and other ions within the intact cell and of making these measurements continuously has caused great interest. In all cases, these electrodes are the result of painstaking development of personal techniques; only an outline of the

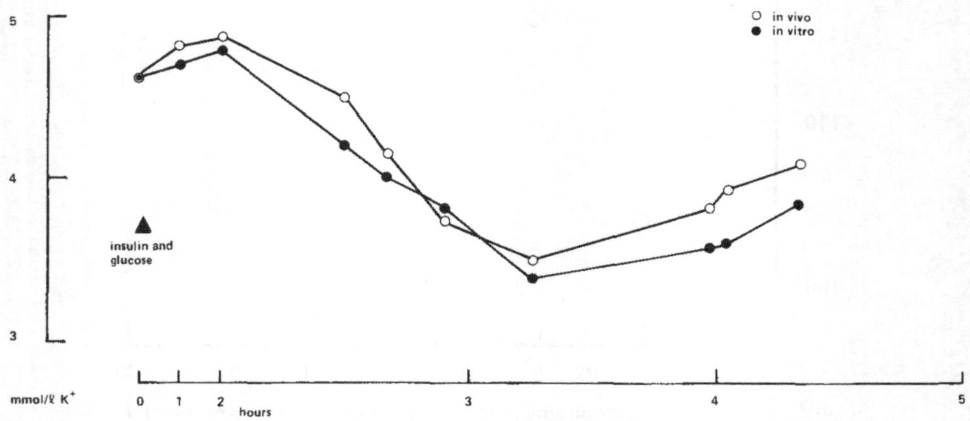

FIGURE 12. At time 0, the catheter electrode is calibrated by withdrawing a blood sample. Thereafter, catheter readings are made continuously, while the potassium changes in response to an injection of insulin and glucose. Results are compared with measurements made on blood samples withdrawn intermittently.

development of microelectrode techniques and supply references to publications by the most active workers will be given here.

Intracellular pH measurements with tungsten microelectrodes were performed by Caldwell,[37] but, later, glass electrodes were used intracellularly.[38,39] Hinke[41] made intracellular measurements with sodium-sensitive glass microelectrodes. These spear-shaped glass electrodes were of high resistance, and the resistance increased as the sensitive tip diminished in length. This meant that only large cells could be studied, and even then there was uncertainty as to whether the full length of the cation-sensitive glass was intracellular. The development by Thomas of recessed-tip electrodes for Na^+,[42,43] and, later, H^+ [44] meant that an electrode with a small-tip diameter could have a relatively large area of ion-sensitive glass exposed to the intracellular contents (Figure 13). Hinke and Menard[45] have recently compared the results of intracellular glass electrode and DMO (5, 5-dimethyloxazolidine - 2, 4-dione) methods of measuring pH and confirmed that there is reasonable agreement.

The development by Walker[46] of a liquid ion-exchanger microelectrode increased the range of measurements that could be made. Intracellular potassium and chloride activities in heart muscles were measured by this technique.[47,48,49] Similar measurements were made in skeletal muscle,[50,51] and the technique has been applied to the study of renal mechanisms by Khuri.[52] Thomas's group has applied the same principle of recessed-tip electrodes in liquid membrane electrodes.[53,54] Khuri extended the range of liquid ion-exchanger microelectrodes with a bicarbonate sensor for renal research.[55] Apart from intracellular use, these liquid ion-exchanger microelectrodes have been used to study extracellular potassium fluxes in neurophysiology by Kriz et al.[56,57] and Lothman et al.[58,59] Potassium fluxes from muscle have been followed with liquid ion-exchanger electrodes.[60,61]

The next stage in the development of macroelectrodes was incorporation of the liquid ion exchanger in a polymer membrane. Miniature electrodes based on polymer

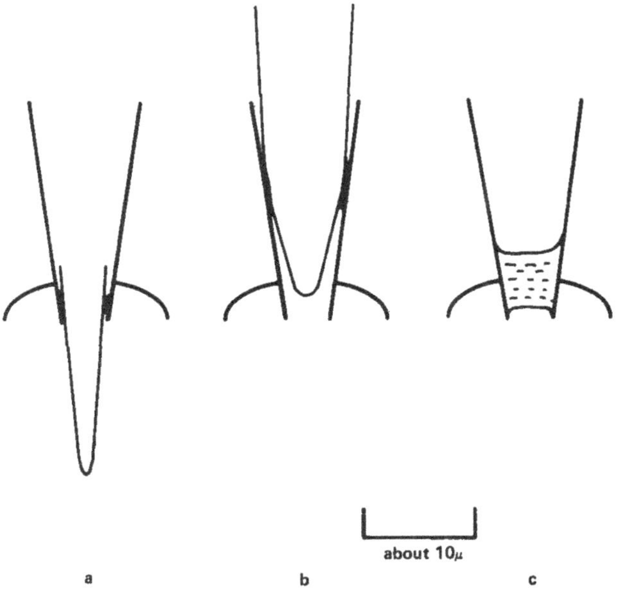

FIGURE 13. Examples of types of microelectrodes. (a), (b). Illustration of the advantage of the recessed-tip design in reducing the cell penetration required to achieve maximum exposure of ion-selective glass to the intracellular contents. (c). A liquid ion-exchanger microelectrode.

membranes are in use. Gardner-Medwin,[61] using a miniature electrode with a PVC membrane cast onto porous ceramic,[35] has studied current-induced potassium fluxes across the brain. Ryba and Petranek[62] have also miniaturized potassium electrodes by mounting PVC membranes on glass.

Microelectrodes therefore usually necessitate the use of micro-reference systems. Micro-glass pipettes are often used either separately or as a double electrode with the sensing electrode. Extremely fine glass pipettes often exhibit spurious "tip potentials" when compared with a conventional liquid junction. The construction and use of these electrodes tend to be an art rather than a science.

It is significant that while these techniques are now well established and more studies are appearing in the physiological literature, they tend to come from relatively few centers. Studies with glass electrodes on the pH and its control in snail neurones,[63] crab muscle,[64] mouse muscle,[65] and mammalian cardiac muscle[66] represent the application in various tissues of difficult but well-tried techniques. Published symposia reports under the editorships of Berman and Hebert[67] and Kessler et al.[68] provide an introduction to the techniques available in this area of biological research.

VIII. CONCLUSIONS

Ion-selective electrodes have very exciting possibilities for research workers, only a few of which have been touched on in this chapter. Inevitably, in this area, trial and error continues to operate, and the individual research worker modifies his analytical tools to suit his own requirements. The much more uniform clinical field requires a different approach.[69] Equipment should be reliable and fully evaluated before being put into medical use. Furthermore, it is of no help to first develop equipment and then look for a place to use it. We hope that by outlining the clinical requirements and the practical limitations of the methods available we have offered a realistic summary of the place of ion-selective electrodes in medicine and medical research.

REFERENCES

1. **Hirst, A. D.,** *Proc. Anal. Div. Chem. Soc.*, 14, 345, 1977.
2. **Dahms, H.,** *Clin. Chem.*, (Winston-Salem, N.C.), 13, 437, 1967.
3. **Smith, M. D., Rogers, R. W., Genshaw, M. A., and Greyson, J.,** *Clin. Chem.*, (Winston-Salem, N.C.), 19, 891, 1973.
4. **Lustgarten, J. A., Wenk, R. E., Byrd, C., and Hall, B.,** *Clin. Chem.*, (Winston-Salem, N.C.), 20, 1217, 1974.
5. **Moore, E. W. and Wilson, D. W.,** *J. Clin. Invest.*, 42, 293, 1963.
6. **Dahms, H., Rock, R., and Seligson, D.,** *Clin. Chem.*, (Winston-Salem, N.C.), 14, 859, 1968.
7. **Mohan, M. S. and Bates, R. G.,** *Clin. Chem.*, (Winston-Salem, N.C.), 21, 864, 1975.
8. **Foster, J. P.,** in *The Plasma Proteins*, Vol. 1, Putnam, W. F., Ed., Academic Press, New York, 1960, 189.
9. **Gerbrandy, J., Hellendoorn, H. B. A., de Vries, L. A., and Van Leeuwen, A. M.,** *Scand. J. Clin. Lab. Invest.*, 10, 272, 1957.
10. **Gerbrandy, J., Van Leeuwen, A. M., Hellendoorn, H. B. A., and de Vries, L. A.,** *Clin. Sci.*, 19, 181, 1960.
11. **Ebden, P.,** Ph.D. thesis, University of London, 1975.
12. **McLean, F. C. and Hasting, A. B.,** *J. Biol. Chem.*, 107, 337, 1934.
13. **Payne, R. B., Little, A. J., Williams, R. B., and Milner, J. R.,** *Br. Med. J.*, 4, 643, 1973.
14. **Husdan, H., Rapoport, A., and Locke, S.,** *Metabolism*, 22, 787, 1973.
15. **Peterson, K. O.,** *Scand. J. Lab. Clin. Invest.*, 30, 191, 1972.

16. Philpot, G. R., Ph.D. thesis, University of London, 1958.
17. Berry, E. M., Gupta, M. M., Turner, S. J., and Burns, R. R., *Br. Med. J.*, 4, 640, 1973.
18. Schwartz, H. D., *Clin. Chem.*, (Winston-Salem, N.C.), 22, 461, 1976.
19. Thomas, J. D. R. and Moody, G. J., *Proc. Anal. Div. Chem. Soc.*, 12, 48, 1975.
20. Linden, R. J. and Norman, J., *J. Physiol.* (London), 215, 491, 1971.
21. Lamb, A. B. and Larson, A. T., *J. Am. Chem. Soc.*, 42, 29, 1920.
22. Band, D. M., Kratochvil, J., and Treasure, T., *J. Physiol.* (London), 265, 5P, 1977.
23. Band, D. M. and Kratochvil, J., *J. Physiol.* (London), 239, 10P, 1974.
24. Leblanc, O. H. and Grubb, W. T., *Anal. Chem.*, 48, 1658, 1976.
25. Covington, A. K. and Robinson, R. A., *Anal. Chim. Acta*, 78, 219, 1975.
26. Fleet, B., Ryan, T. H., and Brand, M. J. D., *Anal. Chem.*, 46, 12, 1974.
27. Friedman, S. M., Jamieson, J. D., Hinke, J. A. M., and Friedman, C. L., *Proc. Soc. Exp. Biol. Med.*, 99, 727, 1958.
28. Friedman, S. M., Jamieson, J. D., Nakashima, M., and Friedman, C. L., *Science*, 130, 1252, 1959.
29. Band, D. M. and Semple, S. J. G., *J. Physiol.* (London), 184, 58P, 1966.
30. Band, D. M. and Semple, S. J. G., *J. Appl. Physiol.*, 22, 854, 1967.
31. Staehelin, H. B., Carlsen, E. N., Hinshaw, D. B., and Smith, L. L., *Am. J. Surg.*, 116, 280, 1968.
32. Filler, R. M., Das, J. B., and Espinosa, H. M., *Surgery*, 72, 23, 1972.
33. Filler, R. M. and Das, J. B., *Advances in Experimental Medicine and Biology*, Vol. 50, Plenum Press, New York, 1974.
34. Leblanc, O. H., Brown, J. F., Klebe, J. F., Nierach, L. W., Slusarczuk, G. M. J., and Stoddard, W. H., *J. Appl. Physiol.*, 40, 644, 1976.
35. Band, D. M. and Treasure, T., *J. Physiol.* (London), 266, 12, 1977.
36. Soutter, L. P., Conway, M. J., and Parker, D., *BioMed. Eng.*, 10, 257, 1975.
37. Caldwell, P. C., *J. Physiol.* (London), 120, 31P, 1953.
38. Caldwell, P. C., *J. Physiol.* (London), 126, 169, 1954.
39. Caldwell, P. C., *J. Physiol.* (London), 142, 22, 1958.
40. Eisenman, G., Rudin, D. O., and Casby, J. V., *Science*, 126, 831, 1957.
41. Hinke, J. A. M., *Nature*, 184, 1257, 1959.
42. Thomas, R. C., *J. Physiol.* (London), 210, 82P, 1970.
43. Thomas, R. C., *J. Physiol.* (London), 220, 55, 1972.
44. Thomas, R. C., *J. Physiol.* (London), 238, 159, 1974.
45. Hinke, J. A. M. and Menard, M. R., *J. Physiol.* (London), 262, 533, 1976.
46. Walker, J. L., *Anal. Chem.*, 43, 89A, 1971.
47. Walker, J. L. and Ladle, R. O., *Am. J. Physiol.*, 225, 263, 1973.
48. Walker, J. L. and Ladle, R. O., *Advances in Experimental Medicine and Biology*, Vol. 50, Plenum Press, New York, 1974, 159.
49. Ladle, R. O. and Walker, J. L., *J. Physiol.* (London), 251, 549, 1975.
50. Armstrong, W. M. and Lee, C. O., *Science*, 171, 413, 1971.
51. Lee, C. O. and Armstrong, W. M., *J. Membr. Biol.*, 15, 331, 1974.
52. Khuri, R. N., Hajjar, J. J., and Agulian, S. K., *J. Appl. Physiol.*, 32, 419, 1972.
53. Neild, T. O. and Thomas, R. C., *J. Physiol.* (London), 231, 7P, 1973.
54. Neild, T. O. and Thomas, R. C., *J. Physiol.* (London), 242, 453, 1974.
55. Khuri, R. N., Bogharrian, K. K., and Agulian, S. K., *Pfluegers Arch.*, 349, 285, 1974.
56. Kriz, N., Sykova, E., Ujec, E., and Vycklicky, L., *J. Physiol.* (London), 238, 1, 1974.
57. Kriz, N., Sykova, E., and Vycklicky, L., *J. Physiol.* (London), 249, 167, 1975.
58. Lothman, E., Lamanna, J., Cordingley, G., Rosenthal, M., and Somjen, G., *Brain Res.*, 88, 15, 1975.
59. Lothman, E. and Somjen, G., *J. Physiol.* (London), 252, 115, 1975.
60. Hnik, P., Kriz, N., Vyskocil, F., Smiesko, V., Mejsnar, J., Ujec, E., and Holas, M., *Pfluegers Arch.*, 338, 177, 1973.
61. Gardner-Medwin, A. R., *J. Physiol.* (London), 269, 32P, 1977.
62. Ryba, O. and Petranek, J., *Talanta*, 23, 158, 1976.
63. Thomas, R. C., *J. Physiol.* (London), 238, 159, 1974.
64. Aickin, C. C. and Thomas, R. C., *J. Physiol.* (London), 252, 803, 1975.
65. Aickin, C. C. and Thomas, R. C., *J. Physiol.* (London), 260, 25P, 1976.
66. Ellis, D. and Thomas, R. C., *J. Physiol.* (London), 262, 755, 1976.
67. Berman, J. H. and Hebert, N. C., *Advances in Experimental Medicine and Biology*, Vol. 50, Plenum Press, New York, 1974.
68. Kessler, M., Clark, L. C., Lubbers, D. W., Silver, I. A., and Simon, W., *Ion and Enzyme Electrodes in Biology and Medicine*, University Park Press, Baltimore, 1976.
69. Treasure, T. and Band, D. M., *Proc. Anal. Div. Chem. Soc.*, 14, 334, 1977.

Chapter 4

ANALYTICAL METHODS INVOLVING ION-SELECTIVE ELECTRODES (INCLUDING FLOW METHODS)

K. Toth, G. Nagy, and E. Pungor

TABLE OF CONTENTS

I. INTRODUCTION

A. Theoretical Considerations

Quantitative chemical analytical methods consist of a series of different individual steps. However, these individual steps are continuously developing with advances in analytical research. Furthermore, the significance of the individual steps may also change, within an analytical method suitable for the solution of a certain individual problem. For example, the analytical value of certain methods may be altered by omitting a sophisticated step of the procedure.

Of all the individual steps of the analysis, detection has a special significance. The analytical evaluation is fundamentally based on the relation between the detector signal and the concentration or amount of the substance to be analyzed. In general, this relation can be expressed as follows:

$$I = f(c_x, c_k, P, T) \tag{1}$$

where I is the signal level, c_x is the unknown concentration, c_k is a constant expressing the effect of the other components also present in the sample (i.e., the so-called "medium effect"), P is the pressure, and T is the temperature. The relation is greatly influenced by the properties of the detector itself.

The recently developed ion-selective electrodes form a group of detectors of almost similar properties among the available analytical tools. In working out a new analytical method suitable for the solution of a certain problem, the characteristics of the problem determine the sequence of the necessary individual steps and influence also the selection of the appropriate kind of detector. At the same time, the properties of a detector exert a basic influence upon the analytical method itself. A detector will be used more widely the simpler it is, i.e., the less restrictions it places on the analytical method applied. The smaller the number of the individual steps (enrichment, separation, addition of reagent, incubation, etc.) needed in the analytical procedure as a requirement of the detector, the wider the application for the method. Detection by means of ion-selective electrodes can, in most cases, be directly adapted, without including any further more complicated individual steps, to many earlier developed and widely used analytical methods, rendering these more simple, rapid, or accurate. At the same time, new analytical methods developed with the help of ion-selective electrodes are very simple and can be performed in any laboratory. The main advantage of ion-selective electrode methods therefore lies in their simplicity.

A short and general survey of detector properties important from the point of view of analytical methods follows. Accordingly, the most important ion-selective electrodes will briefly be surveyed, and it will be shown in Section II how the general properties of ion-selective electrodes influence analytical methods and their individual steps. From the point of view of modern analytical chemistry, selectivity is undoubtedly the most important property of a detector. The selectivity is a complex property of the detector, which manifests itself in the formation of the detector signal by distinguishing among the components present in the multicomponent system; i.e., in the course of signal formation, one of the substance species is preferred.

In the case of different detector types, an appropriate selection of the unit to be used for the characterization of their selectivity is generally not a simple task, since selectivity is a common feature of both the system and the detector. Accordingly, the expressions and methods for the definition or measurement of the selectivity contain, in most cases, concessions and often are exact only in systems consisting of certain components.

For the characterization of the selectivity of ion-selective electrodes, the selectivity coefficient is generally accepted, which is defined by the Nicolsky equation:

$$E = E_0 + \frac{RT}{z_i F} \ln \left(a_i + \sum_{j=i}^{n} k_{ij}\, a_j^{\,zi/zj}\right) \tag{2}$$

where R, T, F, z_i, and a have their conventional electrochemical meanings, i is the component preferred by the electrode, j is one of n other ions present in the system, and k_{ij} is the selectivity coefficient.

As can be seen, according to the definition, the selectivity coefficient of the ion-selective electrode refers to the ion preferred and to another ion present in the system.

From the point of practical analysis, it is advisable to characterize the selectivity of an ion-selective electrode detector with as many data as possible. Unfortunately, for the measurement of the selectivity coefficient of ion-selective electrodes, there exists no generally accepted method. Accordingly, it is rather hard to compare the published data and without knowledge of the measuring method, the data are not of much use. The development of a uniformly accepted and used measuring method is expected in the near future. In our opinion, this method will be the so-called "mixed-solution procedure,"[1] since this cannot yield inaccurate and faulty results even with the less selective electrodes. The other, technically less demanding methods, the so-called "separate-solution procedure," yields data of practical use only in certain cases.

Fortunately, many ion-selective electrodes can be used as detectors because of their favorable selectivities. In many cases, with the help of ion-selective electrodes, the analysis can be performed directly even in complex systems. Due to the favorable selectivity of ion-selective electrode-based detection, analysis can often be simplified by omitting a step for separation or masking. Naturally, even the selectivity of the best ion-selective electrode has its limits. Often, a component occurs in the system to be analyzed which, under the given conditions, influences the detector signal and interferes with the measurement. In such cases, a separation or masking procedure must be inserted at the appropriate place within the series of steps of analysis. However, these steps, as well as the applied reagents, must create favorable conditions for the ion-selective electrode-based detection.

In principle, the analysis of a system of components can be solved with the help of a detector system containing x electrodes of different types with well-defined selectivity factors. However, the successful application of this method of eliminating the interfering effect by means of a multielectrode system is rather limited. The reason for this lies in the lack of exact knowledge of the relationship describing the selectivity coefficient and in the limited number of the ion-selective electrode types.

The second important detector property is measuring range. A favorable concentration region of the detector fundamentally determines the applicability of a given method employing an electrode as detector. Especially at the lower measuring limit, the sensitivity is decisive, since dilution of the sample solution generally does not involve any problem. Detection by means of ion-selective electrodes may be performed over an exceedingly broad concentration range, which, for certain electrodes, may embrace five orders of magnitude.

The lower limit of the concentration range greatly depends not only on the properties governing the electrode function, such as stability constants of complexes, solubility product, distribution quotient, etc., but also on such factors as the nature of the other ions and their concentration, other components present, pH of the solution, and so on. Of special significance is whether the sample solution and the calibrating solutions are buffered for the ion to be measured or not. In buffered systems, the measurement of ions even at very low concentrations is possible. But the measurement of very diluted non-ion-buffered solutions cannot be performed with the help of calibrations carried out in buffered systems.

The lower detection limit, probably for kinetic reasons, is shifted in the direction of lower concentration values if the relation between the electrode surface and sample solution changes, i.e., if the measurement is carried out in mixed or flowing solutions.

Accordingly, the lower limit of ion-selective electrode-based detection can be favorably influenced by suitable selection of the measuring conditions. In order to increase the lower limit of measurement, it is often advisable to remove the interfering ions (by ion exchange, extraction, precipitation, distillation, etc.) or to apply the method of detection worked out for flowing systems.

The lifetime of the detector is the time period during which a given detector operates in a reliable way without renewal or special cleaning. The lifetime of a detector is a very important parameter from the point of view of an analytical application, and it is evident that a long lifetime is desirable. The use of a detector of relatively short lifetime is economically justified only if the applied measuring method allows the analysis of a great number of samples in a relatively short time. However, the applied analytical method itself may often exert an influence upon the lifetime of the detector. Ion-selective electrodes may be long-lifetime sensors. Certain types, e.g., the precipitate-based electrodes and the neutral carrier-based electrodes, conserve their electrochemical properties over several years. Other electrode types, such as the liquid membrane electrodes, gas electrodes (SO_2), and enzyme electrodes, however, operate for only a few weeks in a satisfactory way. The measuring conditions, e.g., aggressive substances present in the sample solution, substances adhering to the electrode, substances dissolving the active material from the surface layer of the electrode, or electrode-damaging components, similarly may shorten the lifetime of the electrode.

To employ electrodes with short lifetimes in analytical practice, the development is necessary of measuring techniques which permit rapid analysis of many samples as well as repeated recalibration of the electrode. An example is the application of the so-called "injection technique" (to be discussed in Section IV, A.3.c), which can advantageously be used with electrodes of short lifetime, e.g., enzyme electrodes.[2]

When the sample solution contains electrode poisoning substances, then it is advisable to use a measuring technique in which the sample solution is in contact with the electrode only for a short time. Furthermore, there is often a possibility of using an indirect method which eliminates the electrode-poisoning effect.

If in an ion-selective electrode detector-based analytical method, one must reckon with a change of the electrode function, and/or a reduction of the electrode sensitivity, then it is very important to check the electrode properties several times during the measurement. Accordingly, in such cases, it is advisable to give, along with the description of an analytical method, a rapid, easily performed technique for checking the electrode properties. It is desirable to provide the user with information on what diminished sensitivity is acceptable and when the electrode should be exchanged or renewed.

Shape, size, and dimensions of the detectors may also exert a significant influence upon their application to analytical methods. Shape, dimensions, and operation of a detector may define the volume of the sample, as well as the rate of sample exchange applicable. Of course, the effect is a mutual one, and therefore, if possible, it is advisable to take into consideration the requirements of the analytical method in the selection of a new detector.

From this point of view, ion-selective electrode detection is very favorable, since properly operating ion-selective electrodes of various shapes, sizes, and structures can be made with the use of materials of different construction. A great variety of electrodes are known; among others, there are needle electrodes with a measuring surface of a few square microns, multichannel needle electrode systems, robust industrial measuring electrodes, ion-selective electrodes with flat or spherical measuring surfaces, and microcapillary electrodes forming a flow-through channel. Enzyme electrodes[3] for the analysis of drop-sized samples, coated-wire electrodes, and flow-through cap-type electrodes are good examples which demonstrate how well the structure of an ion-selective electrode can satisfy the requirements of a special analytical method.

Though several of the specially shaped ion-selective electrodes are commercially available, in the laboratory it is mainly electrodes of conventional shape and dimensions that are used. These can easily be handled and are not sensitive to mechanical effects. With their help, analysis even in sample solutions of a few tenths of a milliliter can be performed.

In the application of various detectors to analytical methods, it is important that a given detector should operate well in the medium required by the method. From this point of view, the following can be said about ion-selective electrodes as detectors: The application of ion-selective electrodes requires ion-dissociating media, i.e., solutions prepared from solvents of high dielectric constant. Knowledge regarding the behavior and applicability of ion-selective electrodes is mainly restricted to aqueous solutions. Experiments performed in mixed aqueous organic solvents by means of ion-selective electrodes are still in an early stage.

It is evident that the possibility of measurements with ion-selective electrodes in organic solvents depends on what extent the solvent may damage, by its dissolving effect, the measuring membrane or other structural unit of the electrode. From this point of view, liquid membrane electrodes are less favorable than precipitate-based, ion-selective electrodes. The presence of organic solvents in the sample may lead to further problems in the case of enzyme and gas electrodes. Components causing turbidity in the solution and aggressive and corrosive materials do not usually interfere with the operation of ion-selective electrodes, however.

Ion-selective electrodes are suitable for the direct analysis of gases. In this application, the measuring surface of the electrode is coated with an aqueous layer of buffer solution to ensure the proper function of the electrode (regarding gas electrodes and air gap electrodes, see Chapter 1, Volume II).

The field of application of a detector may also be influenced by the time required for the measurement. Often, this requirement depends fundamentally on the response time of the sensor, i.e., on the time elapsing before the attainment of the detector signal characteristic of the system after the instant of contact of the system and the detector. It is evident that, on the one hand, with detectors of long response time, rapid processes cannot be followed; on the other hand, if the detector requires a relatively long measuring time, then it is impossible to perform rapid analysis of a great number of samples.

Though the value of the response time may significantly differ, depending on the electrode type, from an analytical point of view, the response time of ion-selective electrodes is usually adequate, and for certain ion-selective electrodes, it is only a few milliseconds. Accordingly, the response time of an electrode is not a significant factor determining the rate of an ion-selective electrode-based measuring technique. The time necessary for the measurement is rather determined by the properties of the whole measuring cell, such as diffusion and convection processes at the boundaries, the structure of the liquid-liquid interphase, its character, and so on.

It is worthwhile to note that in ion-selective electrode-measuring cells, the electromotive force (EMF) values are more rapidly stabilized in a stirred solution. Owing to the high ion activity, this is especially true for buffered solutions. As will be shown later in detail, in flowing systems the development of a stable EMF is more rapid than in mixed or static solutions. In certain methods, by means of an appropriately developed ion-selective electrode sensor, an unbelievably high rate of analysis can be attained.[4,5]

It should be pointed out that response times of ion-selective electrodes or the delay time, hold-up time, and dead time of the ion-selective electrode cells may change in the course of examination. With measurements performed in dynamic systems, for example, with automatic titrations, this may cause errors.

B. Some Practical Devices for Improving Selective Determination

Some examples will be described of procedures for making the selective measurement possible.

1. Procedures for Decreasing Interfering Effects

Cheng and Cheng,[6] using a bivalent cation electrode for the measurement of magnesium, found that the determination can be performed selectively even in the presence of numerous other multivalent cations; if ethylene glycol-*bis*(2-amino ethylether) tetraacetic acid (EGTA) is added to the buffered sample of pH 7, the interfering ions (Ca^{2+}, Zn^{2+}, Pb^{2+}, Ni^{2+}, Co^{2+}, Cd^{2+}, Mn^{2+}, Fe^{2+}, and Fe^{3+}) become masked through complex formation.

This problem arose in the selective determination of the exchangeable calcium and magnesium ion contents in soils by means of ion-selective electrodes. The soil sample was extracted with ammonium acetate; after an hour, the extract was filtered off, washed with ammonium acetate, and dried. The dry residue was dissolved in 0.5 *M* hydrochloric acid and washed with distilled water. The measurement with an ion-selective electrode was carried out after the sample solution had been buffered to pH 7. In addition to the bivalent cation-selective electrode as measuring electrode, a calcium-selective electrode was used.

The cesium ion selectivity of a liquid electrode containing cesium tetraphenyl borate previously dissolved in 4-ethylnitro-benzene is greatly influenced by NH_4^+ and Hg^{2+} ions. Baumann[7] suggested a method for cesium measurement under these conditions, according to which cesium ions are extracted from a strong alkaline solution with 4-*sec*-butyl-2(α-methyl-benzyl)phenol (BAMP) dissolved in cyclohexane, whereupon the cesium ions are reextracted with HCl solution. The determination of cesium ions was carried out in a solution whose pH value was adjusted with Tris buffer solution.

The influence that interfering ions exert upon the operation of ion-selective electrodes can be eliminated with the application of ion exchangers. Hulanicki et al.[8] chose this method to eliminate the disturbing effect of calcium ions in the potentiometric titration of sulfate ions with lead nitrate. To effect exchange of the calcium ions, Amberlite® HP 120 in hydrogen form was used. For the determination of the nitrate ion content of natural waters, Hulanicki and co-workers[9] eliminated interfering effects by adding a suitable buffer solution containing various precipitate-forming ions.

With some enzyme electrodes, the presence of ionic species to which the fundamental electrode responds disturbs the determination. This problem can also be eliminated with the removal of the interfering ions by ion exchange.[10]

The potentiometric titration of SO_4^{2-} ions with lead nitrate using a lead ion-selective indicator electrode is disturbed by Cl^- and HCO_3^- ions. The interfering effect is twofold; chloride ions disturb the electrode response and influence also the titration reaction itself. By the application of suitable ion exchangers, the interfering ions can be removed.[11] For this purpose, the solution is first flowed through a silver-form and then a hydrogen-form cation-exchanger resin column which binds the interfering ions.

Anion-exchanger-based ion-selective electrodes are not highly selective to nitrate ions, and are influenced by numerous other anions present in soils and water samples. The values of the selectivity coefficients of these anions are as follows: NO_3^- (1); Br^-, S^{2-}, NO_2^-, CN^-, and HCO_3^-, (approximately 10^{-2}); $RCOO^-$, Cl^-, and CO_3^{2-} (approximately 10^{-3}); and SO_4^{2-}, $H_2PO_4^-$, and F^- (approximately 10^{-4}). According to Milham et al.,[12] the interfering effect can be decreased by adding an appropriate buffer to the sample solution. The components of the buffer eliminate the interfering effects either by complex formation, by precipitate formation, or otherwise. Besides eliminating the interfering effects, the addition of the buffer solution ensures identical measuring conditions.

Guilbault and Nagy[13] worked out a method for the elimination of the interfering effect of ionic components upon the operation of the urea enzyme electrode by ensuring identical interference level at the calibration and the actual measurement. Thus, the calibration of the enzyme electrode is carried out in standard urea solutions which contain the interfering ions in identical concentration to each other. Sample solutions either are diluted with the inert buffer solution, or a solution containing interfering ions is added to the samples until the fundamental electrode, which does not contain the enzyme reaction layer (in the present case, an ammonium ion-selective electrode), shows an identical electrode potential with that observed in the standard solution. After performing the measurement by means of the enzyme electrode, the urea concentration of the sample solution can be obtained with the help of the calibration curve, considering also the dilution factor. In this way, in the course of the measurement and calibration, the expression involving $\Sigma k_{i,j} \cdot a_j$ in Equation 2, which is the interfering effect, is identical within the limits of measuring accuracy. For the application of the method to other types, besides the enzyme and reference electrodes, an auxiliary electrode of identical type to the fundamental sensor of the enzyme electrode is necessary.

2. Steps Employed in Very Diluted Samples

In order to increase the lower concentration limit of the measurement, it is often advisable to remove the interfering ions (by ion exchange, extraction, precipitation, distillation, etc.) or to apply a method of detection worked out for flowing systems. Also, enrichment of the sample can often usefully precede the analytical procedure carried out by means of ion-selective electrodes. To illustrate this kind of approach, some examples will be given.

Fiedler et al.[14] developed an enrichment method for the analysis of less than 2 m*M* CO_2 in natural water and sewage samples by means of a carbon dioxide electrode. In this method, to 10 ml of the sample solution, the pH of which is adjusted by NaOH and HCl to 11.3, 200 μl of 0.1 *M* lead nitrate is added. The precipitate is then filtered off and placed with the filter paper in the vessel of the carbon dioxide air-gap sensor. The CO_2 liberated by 200 μl of 1.0 *M* sodium hydrogen sulfate is measured. In this way, a hundred-fold enrichment can be achieved.

The capability of the conventional methods using ion-selective electrodes can be increased significantly if the ion-selective electrode-based detection is combined with an enrichment process facilitated by a microdiffusion technique. In this way, Hallsworth and co-workers[15] determined submicrogram amounts (10 pg) of fluoride in mineralized vegetable substances, tooth enamel, bones, and other fluoride-containing biological substances. The fluoride ion content is concentrated in a microdiffusion cell in a thin sodium hydroxide layer. The microdiffusion process of 60 hr duration is carried out in a 60% perchloric acid medium and fluoride ion-free atmosphere at 60°C. If the chloride ion content of the fluoride-containing sample is large, then the microdiffusion is carried out in a solution containing 0.12 *M* Ag_2SO_4 and 60% perchloric acid in order to prevent diffusion of the chloride ion.

Elfers and Decker[16] described an ion-selective electrode-based analytical method, combined with an enrichment process, for the determination of very small amounts of water-soluble fluoride in air and various stack gases. The air sample is flowed through a sodium formate-impregnated filter paper with great speed, whereupon the fluorides, bound by adsorption and chemisorption, were eluted with sodium citrate solution. The fluoride content of the eluate was determined by means of a fluoride-selective electrode.

Hrabeczy-Pall et al.[17] determined the fluoride ion content of air after accumulation by absorption in total ionic strength adjusting buffer (TISAB) solution. The air was pumped through the TISAB solution until the fluoride ion content of the absorption solution reached a given concentration value (10^{-5} *M*). From knowledge of the fluoride ion concentration in the TISAB solution and of the amount of air pumped through, the average fluoride ion content of the air samples was determined.

3. Techniques for Increasing the Lifetime of the Detector

A good example of techniques for increasing the detector lifetime is the method developed in the authors' Institute[2] for the determination of cyanide. As is well known,[18] in the operation of precipitate-based ion-selective cyanide electrodes, the active material of the electrode dissolves. This pernicious effect, which may reduce the lifetime of the sensor electrode in very concentrated cyanide solutions, can be eliminated by applying an indirect method for the measurement of the cyanide ion concentration. The cyanide ions are reacted with a small known excess of silver ions, and the activity of the silver ions is detected.

II. THE IMPORTANCE OF SELECTIVE POTENTIOMETRIC SENSORS IN ANALYTICAL PRACTICE

Only a minority of the ion-selective electrodes developed have been widely accepted in analytical practice. Below, some of the more important electrodes will be mentioned.

The analytical value of a selective sensor is determined by the following factors:

1. The demand for the measuring of a component to be determined by an electrode
2. The efficiency, selectivity, sensitivity, and, in most cases, the response time of the electrode
3. The degree of advantage of the application of the electrode from the point of view of measuring technique and methodology, compared with other known analytical methods. Such advantages are, for instance, the higher accuracy of the measuring method or technique, its simplicity of execution, selective determination without separation, shorter analysis time, eventual on-line applicability, good adaptability to the character of the sample, etc.

For the selective determination of fluoride ions, there was previously no reliable, rapid method.[19] The measurement of fluoride ions is, however, of great importance in several fields of chemical analysis, environmental control, and clinical analysis.

The problem was solved by Frant and Ross,[20] who developed a new electrode prepared from a single crystal of LaF_3 doped with Eu^{2+}. This electrode is suitable for the selective measuring of the concentration of fluoride ions to a lower limit of 10^{-6} mol/ℓ. Its function is influenced, from among the anions, only by the hydroxide ion, but this effect can be eliminated by choosing an appropriate pH value. Interfering cations are hydrogen, iron (III), and aluminum ions, which form complexes with fluoride ions. Their effect can be eliminated by choosing an appropriate pH and auxiliary complexing agents. Due to the favorable electrochemical properties of this electrode, it can be used as a detector also in the analysis of numerous ions other than fluoride.[21,22]

For the selective measurement of potassium, spectroscopic methods (flame photometry and atomic absorption) are in competition with direct potentiometry. Although other types have been used, selective measuring of potassium ions over a wide-range analytical application can only be expected from an electrode based on a neutral ligand, such as valinomycin.[23]

In the majority of sample solutions, the selective measurement of potassium ions can be carried out in the concentration range 10^{-1} to 10^{-5} *M* and when there are 5000 times Na^+ and 20,000 times H^+ ions in excess.

In the clinical, environmental control, and industrial analytical laboratories, there is frequently the need to determine calcium ions. In addition to classical and spectrophotometric methods, there are now ion-selective electrodes suitable for measuring calcium ions, based on electrically charged ion exchangers. However, in our opinion, their success is due rather to the urgent demand for a calcium-measuring electrode than to the really favorable electrochemical characteristics of the electrodes. The electrode measuring Ca^{2+} jointly with Mg^{2+} ions, the so-called "water hardness electrode," has a similar structure and importance and similar electrochemical features. As a result of this need, two different kinds of appropriate calcium ion-selective electrode have been developed recently.[24,25]

In analytical chemistry, there are several applicable methods for measuring iodide ions. Probably, this is the reason that the iodide ion-selective electrode, in spite of its favorable electrochemical properties, gained importance mainly owing to its ability to measure cyanide ions.

The measurement of sulfide ions causes frequent problems, especially in food chemistry and water analysis. From the point of view of measuring technique, the development of the silver sulfide-based sulfide ion-selective electrode significantly contributed to the selective determination of sulfide ions. The sulfide ion-selective electrode is a very sensitive and selective anion-selective electrode. Its sensitivity is ensured by thevery low solubility in water of the silver sulfide precipitate. The selectivity of the sulfide ion is influenced by only a few silver precipitate-forming anions and cations and by a few complex-forming ions. In sewage water analysis and the food industry, the selective and sensitive determination of cyanide ions is extremely important. Furthermore, the accurate determination of the free and total cyanide content is frequently required. Commercial electrodes for cyanide do not differ practically and significantly from the iodide or sulfide electrodes produced by the same firm. Any difference lies in the fact that the dissolution of active silver halide precipitate must be taken into consideration in the course of cyanide measurement; therefore, cyanide electrodes are made of thick, mechanically easily renewable membranes.

As the result of research in environmental control, some electrodes measuring heavy metal ions, e.g., Pb^{2+}, Cd^{2+}, and Hg^{2+} ions, have been developed, containing the sulfide of the appropriate heavy metal ion. In general, it is necessary to add silver sulfide to ensure a good electrode function; however, this decreases the selectivity of electrodes.

The accuracy, sensitivity, and selectivity of potentiometric measurements carried out by means of ion-selective electrodes for Pb^{2+}, Cd^{2+}, and Hg^{2+} developed so far do not attain those of other methods applied to the measurement of heavy metals, e.g., atomic absorption methods and modern polarographic measuring techniques.

The determination of nitrate ions is very important in environmental and pharmaceutical analyses as well as in photo-, soil, and agricultural chemistry. There are very few nitrate determination methods at the disposal of the analyst. Thus, it is no wonder that ion-selective electrodes suitable for the measurement of nitrate ions are widely used, in spite of the fact that nitrate-selective electrodes developed so far can hardly be classed among the best ion-selective electrodes.

In the absence of Na^{+}, K^{+}, and Ag^{+} ions, the cation-selective glass electrode can also be used for the measurement of ammonium ions. However, in most cases, the measurement must be preceded by the separation of the sample from the interfering ions and adjustment of the pH. A significant advance was the development of the ammonium-selective electrode, whose active materials are the antibiotics nonactin and monactin.[26] This electrode makes possible the measurement of ammonium ions in the presence of sodium ions in great excess. Adjustment of pH is necessary only to alter the equilibrium between ammonia and ammonium ions.

The total ammonia nitrogen content of the sample solution can be determined after addition of alkali by means of an ammonia gas electrode. From the nature of the sensor, ionic components present in the sample solution do not interfere. Only volatile amines produce an interference response on the electrode.

The ammonium ion-selective electrode and the ammonia gas electrode individually possesses advantages which justify the application of both electrodes. The character of the problem and the conditions of the measurement help decide which of the electrodes is the more suitable for the solution of a given problem. In many cases, there is no possibility of making the solution sufficiently alkaline for application of the ammonia electrode. The presence of potassium ions in an amount which would interfere with the function of a nonactin NH_4^+ electrode may mean the determination is possible only by means of the gas electrode.

Also, the sulfur dioxide and the carbon dioxide electrodes, which belong to the group called "sensitized electrodes," are very important. Selectivity of these electrodes, in the absence of volatile acids, is very good. In acidic medium, the sulfur dioxide electrode is suitable for the measurement of sulfite ions, while the carbon dioxide electrode makes the determination of carbonate ions possible.

Of the enzyme electrodes, the urea electrode is widely used in analytical practice. This is built upon an ion-selective ammonium ion sensor, whose reaction layer contains the enzyme urease. The basic sensor of the electrode is either an ammonium ion-selective electrode containing nonactin as active material or an ammonia or carbon dioxide gas electrode. In our opinion, the most suitable is the ammonium-selective electrode, which works in that pH range which is optimal from the point of view of enzymatic catalysis. An obstacle to the wider use of this type of electrode lies in the fact that the lifetime of the enzyme, immobilized in the reaction layer, is rather limited. Furthermore, the application of the electrode requires utmost care, since the rate of the reaction, catalyzed by the enzyme, may be influenced by many factors. Accordingly the application of this electrode can only be considered if the calibration is repeated very often. However, due to the great technical advantages involved,[27] the carbamide electrode will soon find a wide application in chemical analysis.

The determination of nitrate ions is very important in environmental and pharmaceutical analyses as well as in photo-, soil, and agricultural chemistry. There are very few nitrate determination methods at the disposal of the analyst. Thus, it is no wonder that ion-selective electrodes suitable for the measurement of nitrate ions are widely used, in spite of the fact that nitrate-selective electrodes developed so far can hardly be classed among the best ion-selective electrodes.

In the absence of Na^+, K^+, and Ag^+ ions, the cation-selective glass electrode can also be used for the measurement of ammonium ions. However, in most cases, the measurement must be preceded by the separation of the sample from the interfering ions and adjustment of the pH. A significant advance was the development of the ammonium-selective electrode, whose active materials are the antibiotics nonactin and monactin.[26] This electrode makes possible the measurement of ammonium ions in the presence of sodium ions in great excess. Adjustment of pH is necessary only to alter the equilibrium between ammonia and ammonium ions.

The total ammonia nitrogen content of the sample solution can be determined after addition of alkali by means of an ammonia gas electrode. From the nature of the sensor, ionic components present in the sample solution do not interfere. Only volatile amines produce an interference response on the electrode.

The ammonium ion-selective electrode and the ammonia gas electrode individually possesses advantages which justify the application of both electrodes. The character of the problem and the conditions of the measurement help decide which of the electrodes is the more suitable for the solution of a given problem. In many cases, there is no possibility of making the solution sufficiently alkaline for application of the ammonia electrode. The presence of potassium ions in an amount which would interfere with the function of a nonactin NH_4^+ electrode may mean the determination is possible only by means of the gas electrode.

Also, the sulfur dioxide and the carbon dioxide electrodes, which belong to the group called "sensitized electrodes," are very important. Selectivity of these electrodes, in the absence of volatile acids, is very good. In acidic medium, the sulfur dioxide electrode is suitable for the measurement of sulfite ions, while the carbon dioxide electrode makes the determination of carbonate ions possible.

Of the enzyme electrodes, the urea electrode is widely used in analytical practice. This is built upon an ion-selective ammonium ion sensor, whose reaction layer contains the enzyme urease. The basic sensor of the electrode is either an ammonium ion-selective electrode containing nonactin as active material or an ammonia or carbon dioxide gas electrode. In our opinion, the most suitable is the ammonium-selective electrode, which works in that pH range which is optimal from the point of view of enzymatic catalysis. An obstacle to the wider use of this type of electrode lies in the fact that the lifetime of the enzyme, immobilized in the reaction layer, is rather limited. Furthermore, the application of the electrode requires utmost care, since the rate of the reaction, catalyzed by the enzyme, may be influenced by many factors. Accordingly the application of this electrode can only be considered if the calibration is repeated very often. However, due to the great technical advantages involved,[27] the carbamide electrode will soon find a wide application in chemical analysis.

III. CLASSIFICATION OF ANALYTICAL METHODS

A detailed survey will be given of the analytical methods employing ion-selective electrodes. For this purpose, the methods have been divided arbitrarily into two groups. Analytical methods other than rate methods have been included in the first group, while those based on measuring reaction-kinetic parameters have been arranged into a second group of rate methods. The size of the first group is much larger than that of the second. Measuring methods included in the first group are not limited to methods in which the measurement is conducted on systems in equilibrium or stationary states. All the methods – titration processes, reaction mixture monitoring, and direct potentiometric determinations – are arranged in this group, in which the evaluation of measurement is not carried out on the basis of a signal depending upon the rate of reaction.

The common features of measuring methods included in the first group are summarized as follows: The instantaneous electromotive force (EMF) of the measuring cell, containing the ion-selective electrode, is characteristic of the instantaneous concentration or activity of a component of the system contacting the electrode. Accordingly, it is supposed that the processes from which the electrode response originates are significantly faster than any reactions taking place during measurements in the system measured. Two variations are distinguished. In one case, the ion-selective electrode is sensitive either to the material to be determined itself or to some other material obtained by chemical transformation from the substance to be determined. These methods are called "methods measuring the primary ion" or "direct methods."

In the second case, the material to be determined is allowed to react with another species, and the concentration of the reagent or the substance arising from reaction is followed by ion-selective electrode detection. These methods are called "reagent measuring methods" or "indirect analytical methods."

It must be noted that this classification of methods cannot be regarded as unambiguous. Obviously, there are measuring methods whose classification is rather disputable. It is quite difficult to assign to one of these groups potentiometric argentimetric titrations, for instance.

In the case of methods employing ion-selective electrodes based on reaction-kinetic parameters, i.e., rate methods, the signal dependent on the rate of chemical reaction is measured; upon this basis, usually from a calibration curve, the parameter to be measured (concentration, activity of ions, or activity of catalyst) can be determined.

Measurement and evaluation can be performed in two different ways. In the so-called "difference methods," the evaluation consists of measuring the EMF at particular reaction times or the difference between the EMF values at two reaction times. In the reaction rate methods, based on a transient signal, the EMF (E) is followed continuously during the reaction, and from the measurement, $^{dE}/_{dt}$ is evaluated at a certain instant of time. Most frequently, the evaluation is carried out for the value of $^{dE}/_{dt}$ obtained by extrapolating $t \rightarrow 0$, corresponding to the reaction rate existing at the moment of starting the reaction.

From the point of view of the species to be measured, both the difference methods and transient signal methods can be divided into two groups. The measurement can be of the concentration of the material transformed during the reaction, or it can be of the activity or concentration of a species influencing the reaction rate.

The application of ion-selective electrodes offers two possibilities for progress in analytical chemistry methods. The application to existing analytical methods of ion-selective electrodes gives considerable advantages in measuring technique, and also new methods based on ion-selective electrodes can be and have been developed. The possibilities in both directions are hardly explored. Under these circumstances, a strict classification and treatment of methods involving ion-selective electrodes is not reasonable. The above mentioned classification is aimed at giving only a loosely guiding principle to the discussion. It is impossible to strive for completeness in this survey when new methods are published daily. The most widely adopted analytical methods using ion-selective electrodes will be described, as well as some interesting, recently developed methods.

IV. DIRECT ANALYTICAL METHODS (OTHER THAN RATE METHODS)

The essential significance of ion-selective electrodes lies in the fact that they are suitable for the selective determination of a variety of different ions. Those ions or materials which can be selectively directly determined with the electrode are called "primary ions" or "primary components." In every case, the motivation of the elaboration of a new type of electrode has been the demand for a sensor suitable for measuring a certain sort of material. This has been true at least since the accidental discovery and spreading of the pH-sensitive glass electrode. Consequently, it can be understood why there are such a great number of methods measuring the primary ion or component among those involving ion-selective electrodes.

It is fortunate if the material to be measured selectively by the ion-selective electrode is present in a free state in the sample, for the task then is simply to measure this. However, a number of materials not directly measurable with ion-selective electrodes can be transformed by quantitative chemical reaction to such a component for which there is already an ion-selective electrode at the analyst's disposal.

In this section, the methods suitable for measuring the primary ions present in the sample will be discussed, and then methods will be surveyed by which certain materials can be converted into components detectable by ion-selective electrodes. The methods are illustrated by practical examples.

A. Direct Methods Based on the Measurement of the Primary Ion

The main advantage of applying ion-selective electrodes, besides the possibility of selective measurements, is the simplicity of the technique, provided there are no complicating factors affecting the EMF established in the measuring cell.

Some of the problems and solutions in connection with the determination of primary ions can be listed:

1. Selectivity of ion-selective electrodes developed up to now is, in many cases, not satisfactory, but with the solution of an appropriate procedure, these deficiencies, in particular, the medium effect, can be minimized.
2. The accuracy following from the potentiometric-measuring technique is often not satisfactory, but with the help of suitable methods, the sensitivity of ion-selective electrodes can be increased significantly.
3. It is necessary that analytical procedures employing ion-selective electrodes should be designed for rapid serial analyses.
4. The monitoring of industrial and natural processes, by measuring the ionic concentrations, requires the development of special measuring methods.
5. With appropriate selection of the measuring method, it is possible to optimize the volume of the sample required with the size of the sensor.

In the following section, some of these already widely used methods and some of the lesser known methods will be described. In the course of our discussion, we shall approach the problem from the methodological side by presenting examples of the advantages of using ion-selective electrodes.

1. Direct Potentiometric Method Employing Single Calibration Graphs

The method employing single calibration graphs is the most widely applied potentiometric procedure. Actually, pH measurement falls in this category in which the calibration of the measuring system takes place with the help of two or more standard buffer solutions. The procedure of direct potentiometry is well known, but it is perhaps not superfluous to mention several general points in connection with its application:

1. The potential of the electrode is defined by the activity of the primary ion to be measured. If concentrations are required, then constancy of ionic strength must be ensured during the calibration and the effective measurement as well.
2. In measuring cells with transference, the effect of changes in liquid-liquid junction potential must be kept to a minimum. Therefore, in the course of calibration as well as in the measurement of the sample solution, conditions must be established that such residual liquid junction potentials are small and stable.
3. Salt bridges should not be allowed to contaminate the sample solution and interfere with the electrode function.
4. An error of 1 mV in the measurement of the EMF involves, in concentration measurement, 4% error if univalent ions are involved and 8% error if divalent ions are to be determined.
5. For measurements of high accuracy, it is advisable to thermostat the measuring cell.
6. Extra care is required in the selection of the quality of the electrolyte suitable for the adjustment of ionic strength.

In analytical practice, the potentiometric measuring cell is calibrated with the help of appropriate standard solutions, and the evaluation is conducted using a calibration curve obtained in this way.

In the case of measurements performed with ion-selective electrodes, in contrast to pH measurements, the calibration is generally carried out in systems not buffered to the primary ion.

However, in the literature, the development of metal ion buffers based on complex equilibria applicable for the calibration of some cation-selective electrodes has been described[28-30] (see Chapter 4, Volume I). In practice, these buffer systems appear not to have been widely used, although they have had an important role in electrode appraisal. It is to be expected that ion buffers will increasingly be developed and used in practice, but for a few ions, the development of ion buffers of suitable capacity cannot be expected. For the present, calibration is performed in the different solutions of the well-dissociated salt of the ion to be measured.

As mentioned earlier, particular advantage derives from carrying out the calibration and analysis in the presence of a background solution consisting of appropriate components at high concentration. This background solution ensures optimal conditions for the selective determination of the ion to be measured. The role of background solution is as follows:

1. A constant ionic strength is ensured, as required for concentration measurement.
2. The pH of the sample is adjusted to a value optimal for the measurement.
3. Components disturbing the measurement of the primary ion are complexed or removed by precipitate formation, so that the interfering effects deriving from the presence of ionic components are eliminated.
4. The ratio between the complexed and free forms of the primary ion is kept constant.
5. In calibration and sample solutions, favorable redox conditions are ensured, and faults deriving from the oxidation of components of sample are eliminated.

The background solution is generally applied in the form of a previously prepared stock solution of appropriate concentration is added both to the standard and sample solutions in the required amount. Some background electrolytes of certain compositions applicable in measurements of particular samples have gained widespread acceptance and use. The most commonly employed background electrolyte solutions are those used in fluoride measurements, the so-called "TISAB solutions." Three TISAB solutions of different composition have been proposed (see Table 1). All three solutions consist of ionic strength adjusting salt (NaOH, $NaNO_3$), complexing ligand, and pH buffer. From the three TISAB solutions, the most suitable should be chosen by considering the interfering components present in the sample.[34] In fluoride measurements, when aluminum ions are present in great quantity, it is expedient to use TISAB III, while at high iron (III) and silicate concentration, TISAB II solution is recommended. TISAB I, developed by Frant and Ross,[31] is recommended for use in fluoride measurements when magnesium, calcium, chloride, nitrate, sulfate, and phosphate are present at high concentration.

For the ion-selective measurement of copper (II), a complexing antioxidant buffer (CAB) is advised by Smith and Manahan,[35] the composition of which is as follows: 0.05 *M* sodium acetate, 0.05 *M* acetic acid, 2.0×10^{-2} *M* sodium fluoride, and 2.0×10^{-3} *M* formaldehyde. The buffer is added to the sample in the ratio of 1:1 prior to the potentiometric measurement. This buffer ensures an optimal pH value (pH = 4.8) for copper (*II*) measurement. The acetate ions, present as complexing agent in excess, have a double role. In part, they keep the ratio of the fixed and the free copper ions

TABLE 1

TISAB Solutions

TISAB I[31]
- 500 ml Water
- 57 ml Glacial acetic acid
- 58 g Sodium chloride
- 0.3 g Sodium citrate dihydrate
- (Dilute to 1 liter)

TISAB II[32]
- 500 ml Water
- 170 g (or 85 g) Sodium nitrate
- 68 g Sodium acetate trihydrate
- 92.4 g Sodium citrate dihydrate
- (Dilute to 1 liter)

TISAB III[33]
- 17.65 g Diaminocyclohexanetetraacetic acid (DCTA)
- 500 ml Water
- 40% NaOH, drop by drop until the salt is dissolved
- 300 g Sodium citrate dihydrate
- 60 g NaCl
- (Dilute to 1 liter)

at a constant value; in part, they impede absorption on the wall of the vessel. Also, there is no loss of copper (II) ions if Fe^{3+} ions are present, which could lead to the formation of hydrated iron (III) oxide precipitate occurring with copper (II) coprecipitation. Formaldehyde is the antioxidant component of the CAB solution, but ascorbic acid is also used for this purpose.

For calcium determination, Hulanicki and Trojanowicz[36] suggest a complex-forming buffer (CCB) of the following composition: one liter solution containing 40.4 g potassium nitrate, 3.6 g disodium iminodiacetate, 160 ml 0.5 *M* aqueous acetylacetone solution, 2 ml 10 *M* ammonia, and 1.07 g ammonium chloride. The acetylacetone masks the magnesium, the KNO_3 (0.4 *M*) adjusts the ionic strength, and the iminodiacetate (0.02 *M*) serves to complex the calcium. In biological samples, TEA (1 *M* triethanolamine) background solution can advantageously be applied to calcium measurement.

For the selective measurement of sulfide ions, it is essential to prevent the oxidation of the sample. Therefore, the main component of sulfide antioxidant buffer (SAOB) solution used[37] for the sulfide measurement is the antioxidant, ascorbic acid. SAOB solution is prepared with a glycine-sodium hydroxide buffer for adjusting the ionic strength; its pH is 13.5

When preparing the calibration curve, the standards are chosen so that the concentration of the sample should be in the range covered by the standards. It is expedient to repeat the calibration three times at least, first measuring more and more diluted standards, then taking them in the opposite order toward the more highly concentrated standard solutions; finally, the first step is repeated. The calibration curve is prepared by averaging and plotting the measured values against the concentration or calculating the parameters of the curve by regression analysis to reduce inaccuracy deriving from the drift or hysteresis. Besides the most generally used method of step-by-step calibration carried out with series of single standards, other calibration methods having different specific advantages can be used. From among these methods, we will discuss those which successfully eliminate inaccuracies arising from environmental effects in consequence of solution interchange, e.g., change in liquid junction potential; contamination of standards due to electrode transfer; and change of the electrode surface due to mechanical effects caused by washing and drying.

In each of the three calibration methods mentioned here, the EMF measurements needed to obtain the whole calibration curve are performed in the same measuring cell, generally in continuously stirred solutions.

The so-called "liter-beaker method"[38] is based on the application of a measuring cell of great volume (about one liter). In the calibration, the measuring and the reference electrode are placed into the measuring cell containing the appropriate electrolyte. The concentrated standard solution in increasing volume increments is added to the solution, then, after each mixing, the value of EMF is measured. The calibration curve is plotted on the basis of the concentration calculated, considering the dilution.

According to our experience, electrodes kept out of use for a long time often show uncertain functioning in dilute solutions. Thus, it is not advisable to begin the calibration with dilute solutions, especially when only one series of calibration measurements is carried out. This is the disadvantage of the liter-beaker method.

In a variation of the liter-beaker method, the calibration standards are produced so that the ion giving the electrode response is prepared coulometrically.[39] The advantage of the method lies in the fact that, in addition to the high accuracy of the well-regulable coulometric reagent production, there is no change in the solution; coulometric liter-beaker calibration can be performed in a measuring cell of small volume also. The widespread adoption of the method is hindered by two facts. On the one hand, there are only a small number of ions which can be coulometrically produced with 100% current efficiency in favorable media when measuring with ion-selective electrodes. On the other hand, this method of calibration requires more complicated instruments (potentiostat, integrator, etc.). Furthermore, inaccuracy in this case can also be caused by starting the calibration with the most dilute solution.

A new, continuous calibration process[40] will now be described in detail. The measuring cell in this method is a vessel of constant volume supplied with a stirrer and inflow and overflow channels, called a "first-order container." The apparatus needed for the calibration is presented schematically in Figure 1. Prior to the calibration, the whole vessel is filled with the standard solution of the highest concentration intended to be measured. Then, with continuous monitoring of the EMF of the cell, inert background electrolyte (TISAB, CAB, SAOB, etc.) is brought into the cell at a constant volumetric rate. As a consequence, solution leaves the cell at the same rate. In this way, the concentration of the component giving the electrode response is gradually changing with time. The relationship between the concentration (c) and time (t) is given by

$$c = c_o e^{-\frac{vt}{W}} \tag{3}$$

where c_o is the initial concentration, v the rate of volume flow, and W the volume of the container (constant). From this

$$\log c = \log c_o - \frac{vt}{2.303W} \tag{4}$$

As can be seen, if a constant volumetric rate is maintained, then a linearity exists between the logarithm of the concentration of the solution in cell and the time elapsed since starting the dilution. So the E vs. t values recorded during the continuous calibration without changing the ionic strength of the solution in the measuring cell can be

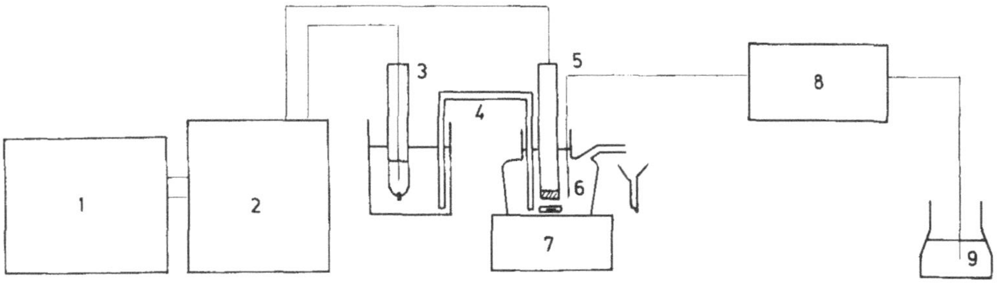

FIGURE 1. Schematic diagram of the apparatus used for continuous calibration: 1, recording apparatus; 2, pH - mV meter or electrometer; 3, reference electrode; 4, electrolyte bridge; 5, indicator electrode; 6, cell with overflow; 7, magnetic stirrer; 8, peristaltic pump; 9, flask containing the "washing"solution. (From Horvai, G., Toth, K., and Pungor, E., *Anal. Chim. Acta,* 82, 45, 1976. With permission.)

simply converted into a calibration curve of E vs. log c by appropriate transformation of the scale of the t axis. A calibration curve of iodide recorded by continuous calibration is compared to that made conventionally in Figure 2.

The advantage of the above method is its simplicity, rapidity, and ease of automation. Since the calibration starts from the more concentrated solutions, proceeding to the weaker ones, this method can be considered as the reverse process of the liter-beaker method. It is free of the above mentioned faults. Due to the character of the method, it can be used especially well for calibrating flow-through potentiometric-measuring cells. The precision of the method, however, depends on how accurately the flow rate is constant. It is obvious that the continuous calibration method can be used only if the response of the electrode is sufficiently fast to follow reliably the concentration change occurring in the cell.

The method of using an ion-selective electrode and doing the evaluation on the basis of a calibration curve will now be described, with an analytical method as an example.

A method has been developed by Milham and co-workers[12] for the determination of nitrate content in plants, soils, and water samples, in which the nitrate ion-selective electrode is used. Both the calibration standards and sample solutions are diluted in ratio 1:1 with an appropriate buffer, the composition of which is as follows: 0.010 *M* aluminum sulfate, 0.010 *M* silver sulfate, 0.020 *M* boric acid, and 0.020 *M* sulfamic acid. The pH of the buffer is previously adjusted to 3 by adding 0.1 *M* sulfuric acid. The measurement was carried out in the range of 5 to 1000 ppm nitrate-nitrogen. In the measuring cell, the salt-bridge electrolyte of the reference electrode contained sulfate, since the conventionally used chloride or nitrate ion-containing bridge would have disturbed the analytical determination.

Preparation of samples is as follows: (a) a pulverized vegetable sample (approximately 100 mg) is dissolved by stirring for 4 min in a mixture of 5 ml distilled water and 5 ml buffer. (b) 20 g of the homogeneous, sieved sample of soil is extracted by rapid stirring for 30 sec in 40 ml water. The mixture is kept at rest for 15 sec; afterwards, it is shaken for 5 sec. The last two steps of extraction are repeated three times. The suspension is allowed to settle, and then the supernatant layer is poured off. The soil extract is diluted in 1:1 ratio with buffer, and the nitrate concentration is determined, (c) a supernatant water sample is diluted with buffer in ratio 1:1, and the nitrate concentration is determined. In determining the nitrate ion concentration, the calibrated electrode pair is placed into the mixed, buffered sample solution and the EMF of the cell determined. The concentration corresponding to the cell potential is determined on the basis of calibration curve.

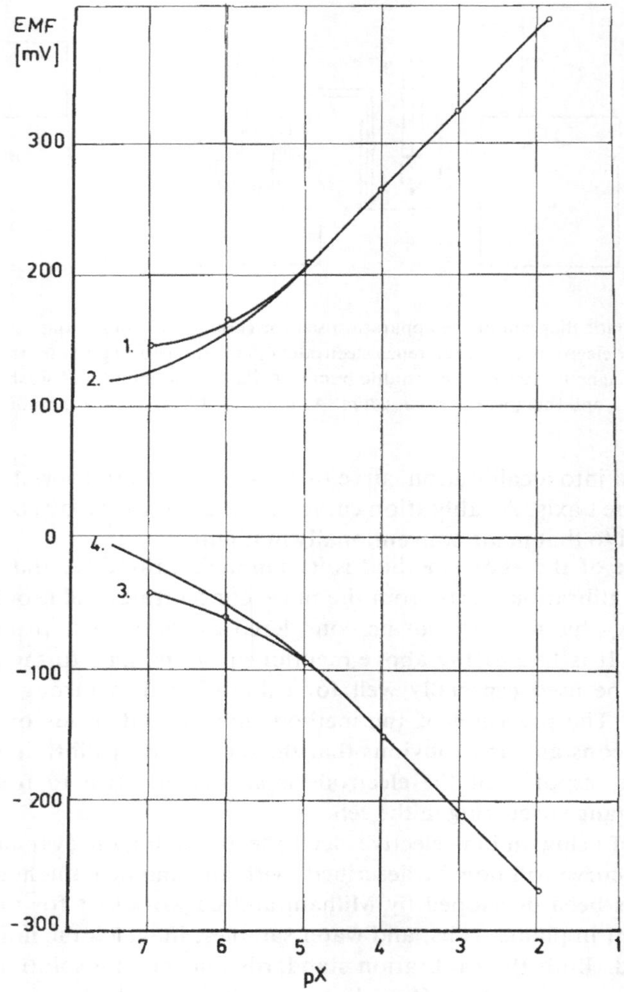

FIGURE 2. Calibration graphs for a silver iodide-based silicone rubber ion-selective electrode obtained by the conventional (Curves 1 and 3) and the continuous (Curves 2 and 4) technique. pX = − log c_{Ag^+} for Curves 1 and 2, while pX = − log c_{I^-} for Curves 3 and 4. (From Horvai, G., Tóth, K., and Pungor, E., *Anal. Chim. Acta*, 82, 45, 1976. With permission.)

This method of measuring nitrate content can be carried out simply with generally available instruments and vessels of analytical laboratories. In some cases, however, even without any fundamental change in the potentiometric measurement, either the character of sample to be measured or the process of preparing the standards requires special laboratory equipment. One such case is illustrated by the following method, which is suitable for the direct measurement of another primary ion. To measure the fluoride content of air, which is very important from the point of view of environmental protection of aluminum factories, some special problems have to be solved:

1. The absorption of fluoride component present in the air sample should be reproducibly ensured in such a way that the fluoride concentration of the absorption liquid falls into the concentration range measurable with the ion-selective electrode.

2. Glass vessels are not suitable for storing the sample and calibration solutions.
3. Gaseous samples can be analyzed accurately only by using gaseous standards, so that both sample and standard are treated in the same way.

A method for the measurement of fluoride content of air, applicable in the permitted emission range, has been developed by Pungor and co-workers.[17] The absorption is performed in a recirculating absorber-distributor unit (see Figure 3) made of PVC or plexiglass by pumping the air through it at a constant rate. The gaseous standards are produced by thermal decomposition of accurately measured quantities of potassium hydrogen fluoride placed in a platinum boat covered by a platinum tube (Figure 3). The concentrations of the standards are determined on the basis of weight decrease. The absorption and measurement are carried out in TISAB I solution (see Table 1). A flow-through measuring cell is used for the potentiometric measurement (Figure 4). For both standards and gaseous samples, the absorbent having the appropriate fluoride concentration is passed into the measuring cell by turning the valve of the distributor. Evaluation of results is done from the calibration curve pertaining to the streaming rate of sample and standard, and absorption time. On the basis of this method, an automated monitor suitable for measuring the fluoride content of air has been developed.

2. Methods Applying Addition Techniques

The second group of measuring methods for primary ion determination is that in which the measurement and evaluation of results are carried out by addition techniques.

The principles of standard addition, subtraction, and sample addition and subtraction methods are dealt with in detail in Chapter 3, Volume I of this book. Here, only some methodological considerations are included as an aid to choosing the proper

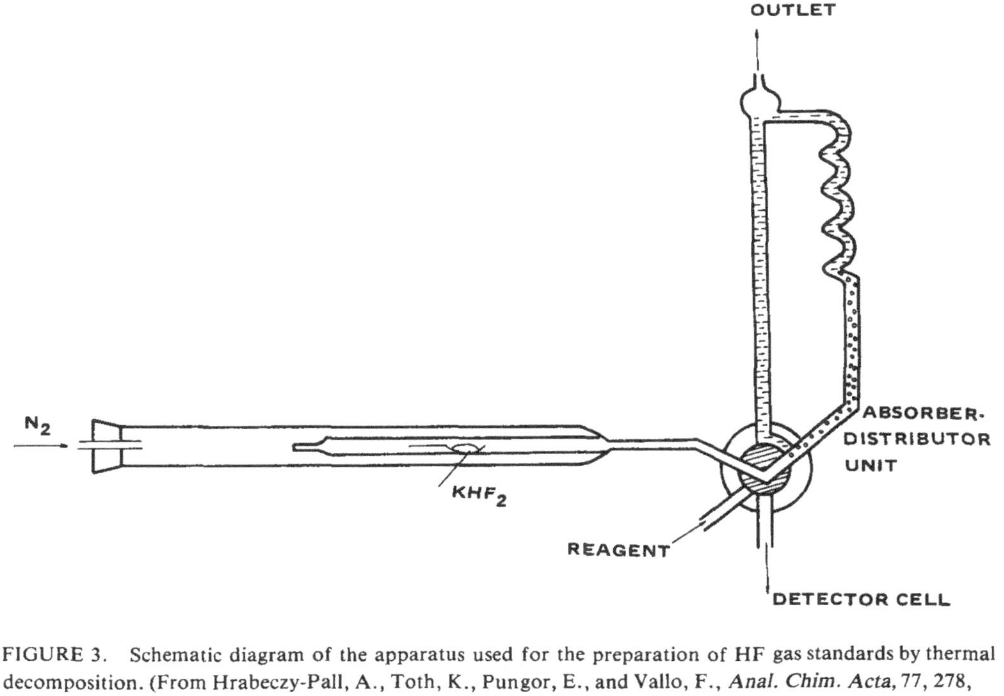

FIGURE 3. Schematic diagram of the apparatus used for the preparation of HF gas standards by thermal decomposition. (From Hrabeczy-Pall, A., Toth, K., Pungor, E., and Vallo, F., *Anal. Chim. Acta*, 77, 278, 1975. With permission.)

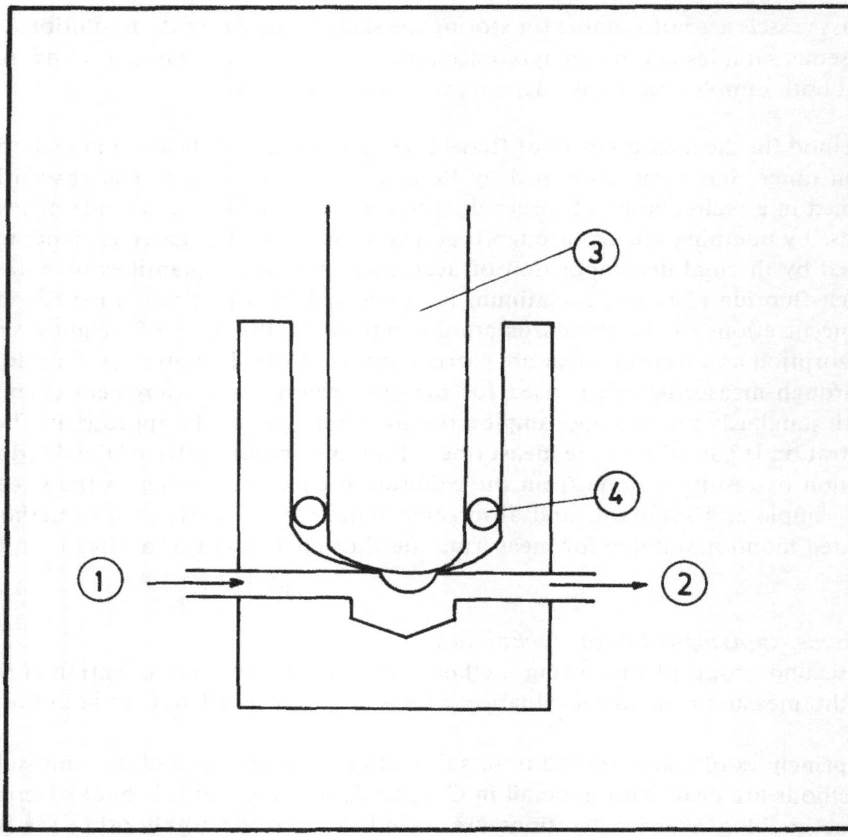

FIGURE 4. Flow-through cell: 1, sample inlet; 2, sample outlet; 3, ion-selective electrode; 4, O-ring. (From Hrabeczy-Pall, A., Toth, K., Pungor, E., and Vallo, F., *Anal. Chim. Acta*, 77, 278, 1975. With permission.)

evaluation method. Later, the applicability of the technique will be demonstrated through a practical example.

Addition methods present two advantages in practice. Both the time and labor of the analysis can be reduced, and errors caused by mechanical effects during the calibration and sample analysis and by the medium effect are expected to be eliminated.

In the application of addition and subtraction techniques, favorable results are ensured only if the following conditions are fulfilled during the addition:

1. The liquid junction potential of the reference electrode bridge solution does not change markedly.
2. The activity coefficient of the component to be determined is kept at a constant value.
3. The ratio of complexes having different stoichiometry from the component to be determined is kept constant.
4. The ion-selective electrode shows a theoretical Nernstian slope in the concentration range examined.
5. Components disturbing the operation of the ion-selective electrode are absent from the system.
6. In the case of subtraction, the concentration following the subtraction is in the Nernstian electrode response range.

Problems arise from the so-called "medium effect," which can more or less be eliminated by means of addition techniques. For example, some surface-active materials give no electrode response, but large organic molecules being present in high, variable concentration may exert an effect upon the EMF through the change of activity coefficient or liquid junction potential.

It is not accidental that the addition technique is the most widely used method in the analysis of biological and environmental samples. According to the addition method, the measurement is performed so that the accurately measured sample solution of volume (V_x) and of unknown (c_x) concentration, which contains the background electrolyte needed for the concentration measurement in an appropriate concentration, is introduced into the measuring cell containing the ion-selective electrode; the EMF is measured (E_x). Afterwards, a standard solution of volume (V_s) and known concentration (c_s) is added to the sample. After appropriate time for mixing, the EMF (E_s) of the cell is measured again. It is reasonable to choose the quantity and concentration of the standard solution so that, due to the addition, the concentration of the ion to be measured is doubled in the cell. The original concentration of the sample solution can be calculated with the following relationship:

$$c_x = \frac{c_s}{10^{(E_s - E_x)/S}\left(1 + \frac{Vx}{Vs}\right) - \frac{Vx}{Vs}} \quad (5)$$

where S is the slope of the electrode calibration curve.

The accuracy of the standard addition measuring technique is somewhat increased if multiple addition is performed. It is then possible to find the slope of the calibration curve of the electrode. However, measurements by multistandard addition have two obvious disadvantages. Multistandard addition increases the time taken for the determination, but also the slope of the calibration curve obtained by multistandard addition relates only to concentrations higher than that of the sample. In the range of low ionic activity, where the electrode response is not linear and the slope is the function of concentration, the combination of the single-known standard addition and slope-by-dilution methods can be advantageously applied. This procedure is especially applicable in clinical and environmental protection analysis; therefore, an example will be given from this area, as developed by Fuchs and co-workers[41] for fluoride measurement in plasma.

In the first step of the method, the potential (E_X) of the properly washed and dried ion-selective fluoride electrode was measured in a 1000-μl plasma sample of unknown fluoride concentration, to which 5% TISAB III solution had previously been added. After removing 100 μl of the sample, which was replaced by the addition of five- to tenfold more concentrated standard solution (100 μl $= V_s$), the EMF (E_2) was measured for this solution. To 500 μl of the above solution, distilled water containing 5% TISAB III was then added to dilute it to half concentration (500 μl), and the EME(E_5) was measured (E_3) in this solution.

On the basis of the equation

$$S = \frac{E_2 - E_3}{\log 2} \quad (6)$$

the slope (S) in the concentration range near the concentration of the sample is determined. The fluoride concentration (c_x) of the plasma can be calculated by the help of the following standard addition equation valid under the conditions of the measurement:

$$c_x = \frac{c_s \cdot V_s}{10^{(E_2 - E_x)/S}(V_s + V_x) - V_x} \quad (7)$$

where V_s is the volume of the standard solution (100 μl), V_x is the volume of the sample (900 μl), and c_s is the concentration of the standard solution.

The combination of the single-known addition and slope-by-dilution methods gives a simple and rapid method which can advantageously be used if the addition, dilution, and the three EMF measurements are carried out in the same vessel, without removing the solution.

An infinite number of other combinations of standard addition, subtraction, and dilution methods can be imagined, and a very large number of such combinations have already been described and applied. From the point of view of general practice, however, the single-known addition and slope-by-dilution combination should be regarded as the method most likely to gain widespread application in the future. The considerable increase in reliability and accuracy is obtained without the complication of the direct potentiometric measuring method.

Recently, more and more significance has been attached to Gran plot methods[42] in the evaluation of measurements performed with ion-selective electrodes.

In the course of standard addition measurement, a graphical evaluation method can be applied as follows. The sample solution appropriately diluted by the basic electrolyte (TISAB, CAB, etc.) with volume V_x is introduced into the measuring cell and the EMF measured. Then, doses of known quantity (V_1, V_2, ...) of standard solution containing the ion to be determined in known concentration (c_s) are added, and, after mixing, the EMF values are measured. Thus, in effect, a multistandard addition measurement is performed. It is expedient to choose the quantity of the primary ion present in the standards so that the difference between the least and highest EMF values should fall in the range $\frac{20-50}{z}$mV where z is the charge on the primary ion. At least five individual standard doses should be applied in succession. After this, the EMF measured vs. standard addition volume is plotted according to Gran's procedure. The points should lie along a straight line. The original primary ion concentration of the sample solution can be calculated from the apparent standard volume given by extrapolating the line to zero.

Gran's plot is a method of linearizing the relation between the primary ion activity, concentration and the EMF measured. The Gran transformation consists of expressing the Nernst equation

$$E = E_o + S \log a \approx E_o + S \log \frac{c_o V_o + cV}{V_o + V} \quad (8)$$

in the form

$$10^{E/S} = \text{constant} \cdot a \approx \text{constant} \; \frac{c_o V_o + cV}{V_o + V} \quad (9)$$

In a Gran plot, the ordinate is $10^{E/S}$, and the abscissa is the quantity proportional to the activity or concentration. Special Gran plot paper (antilogarithmic linear paper) is commercially available, with variations which allow for the effect of volume change. Westcott[43] has prepared a Gran ruler (Figure 5), which has antilogarithmic (E/S) scales for monovalent and divalent ions with volumetric correction. With the help of this ruler, Gran plots can conveniently be done on ordinary millimeter paper (both titrations and multistandard addition or subtraction measurements). A special antilogarithmic converter developed by Van der Meer and Smit[44] can be used to provide a linearized analogue output signal on a millivolt meter:

$$V_{out} = \text{constant} \times 10^{\Delta E/S} \tag{10}$$

The application of the analate addition method is suggested if the volume of the sample to be analyzed is so small that it cannot be measured by the standard addition technique. First, the standard solution of known volume and concentration is poured into the measuring cell. The EMF value is measured; then, after adding known volumes of the sample solution (multianalate addition), containing the same electroactive ions as the standard solution, the value of the EMF is again measured. The unknown concentration is determined with the help of the calculation or the graphical method mentioned earlier. Obviously, the analate addition technique is unable to eliminate the medium effect as a source of error.

In the course of the standard subtraction technique, a known amount of a component which gives no electrode response but quantitatively reacts with the primary ion is added to a known volume of the sample solution, and the variation of the EMF is measured. The unknown concentration is calculated by considering the stoichiometry of reaction or is determined graphically. The use of a multisubtraction method improves the accuracy.

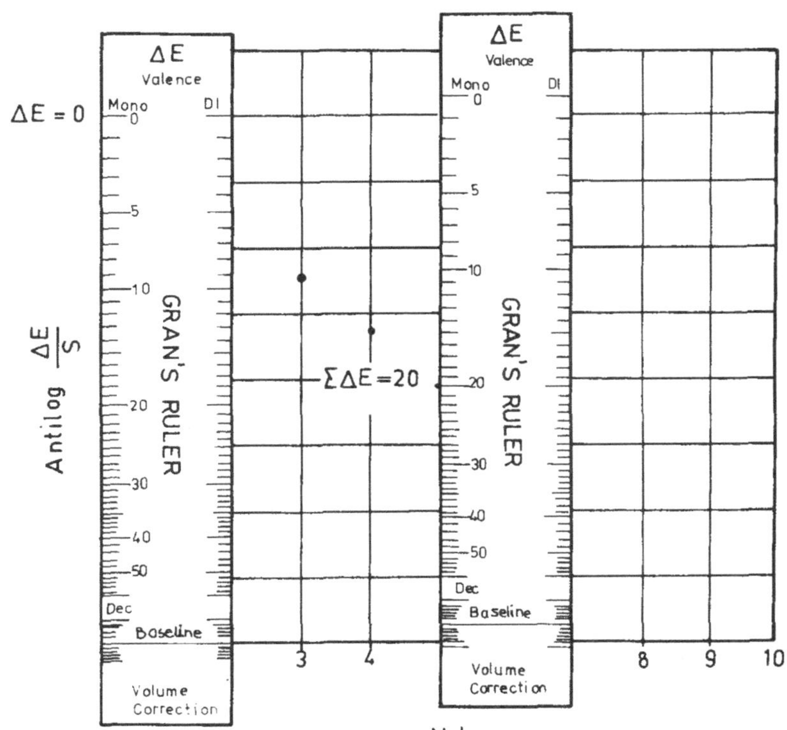

FIGURE 5. The Gran ruler. (From Westcott, C. C., *Anal. Chim. Acta*, 86, 269, 1976. With permission.)

Potentiometric titrations play a prominent role in analytical methods employing ion-selective electrodes (see Volume I, Chapter 3, I.G.).

According to the classification made in this chapter, those groups applying potentiometric titration can unambiguously be listed among methods measuring directly the primary ion only if the reagent (titrant) does not give an electrode response. Consider, for example fluoride concentration determined by precipitation titration with the ion-selective fluoride electrode and lanthanum nitrate titrant solution. If, after having reached the equivalent point, the electrode is regarded as sensitive to the ion activity of the titrant (but see Volume I, Chapter 9, IX. B.), then the titration process can be regarded as a combination of an analytical method based on reagent detection and methods based on direct sample detection. Another example is those types of argentimetric determinations in which a silver halide precipitate-based electrode is employed as indicator electrode. For the evaluation of results, in addition to the conventional graphical or derivative equivalence point location methods, Gran plot methods can be used.

The potentiometric-measuring methods listed above are used not only for direct primary ion determination but also as component steps in other kinds of analytical methods. The measuring method to be selected and applied in combination with the analytical method of optimal performance must be considered separately in each case. The criteria for selection are as follows:

1. Accuracy required
2. Time required in relation to the total time for analysis
3. Quantity and type of interfering ions present in the sample
4. Electrochemical properties of sensor electrode

Two examples will be presented which illustrate the important role of the character of the analytical problem in selecting the most suitable ion-selective electrode-based measuring technique. In our laboratory, there was a demand for a method for the determination of the isopropionamide iodide (N-(3, carbomoyl-3,3-di-phenylpropyl)-N,N-diisopropylmethyl ammonium iodide) active ingredient of a medical suppository containing Probon® (Chinoin Hungarian Pharmaceutical Works, Budapest). The isopropionamide iodide, through its dissociable iodide N-(3,carbomoyl-3,3-diphenylpropyl)-N,N-diisopropylmethyl ammonium iodide or isopropionamide iodide. The isopropionamide iodide, through its dissociable iodide content, can be determined by means of an ion-selective iodide electrode; but in the aqueous solution obtained from the pharmaceutical sample, there was, besides isopropionamide, also Probon® present in a 40 times greater excess.

Due to the possible medium effect, it was not advisable to apply direct potentiometry. Since the argentimetric titration is a very time-consuming technique and a method suitable for the analysis of a great number of samples was required, we carried out comparative investigations regarding the accuracy of different direct potentiometric calibrations, addition techniques, and titration methods.

It was found that when calibration was carried out with a series of standards not containing Probon® the evaluation could not be performed with the necessary accuracy of ± 5%. Accuracy of the measuring technique was significantly improved if the calibration was performed with a series of solutions containing Probon® of a concentration identical with the average sample. By the application of the addition measuring technique with standard dosage, we were successful in obtaining a standard deviation identical to that of the titration method.

Accordingly, determination of the active ingredient (isopropionamide iodide) of the pharmaceutical product was performed as follows. From the medical suppository of known weight, the support material was removed by extraction at 40°C with two samples of 20.0 ml petroleum ether in 30 min. The active material was filtered off and the suspension dissolved in 50.0 ml distilled water. Then, 10 ml of this stock solution was poured into the measuring cell and the EMF of the iodide electrode measured against a saturated calomel electrode. Finally, 100-μl portions of the standard solution of isopropionamide iodide of concentration 1 g/100 ml were added to the cell. The EMF measurement and calculations were performed in the way already described.

The second example selected for the description is a method for following the rate of hydrolysis of dimethyl dichlorovinyl phosphate. This substance contains organically bound chlorine to which an ion-selective chloride electrode does not respond until hydrolysis produces chloride ions. Thus, the rate of the hydrolysis can be followed by measurement of the changes in activity of the chloride ions. In selecting the measuring technique, the titration method cannot be considered, since titrants such as $AgNO_3$ and $Hg(NO_3)_2$ may influence the equilibrium conditions of the autohydrolysis. An organic component whose composition changes with the progress of the reaction will cause a significant medium effect, which can hardly be eliminated by appropriate calibration, and will disturb the evaluation from a calibration curve. The standard addition technique was considered as the only suitable one to follow the process successfully.

The ion-selective electrode measuring techniques, at least in their most generally known versions so far discussed, are connected with EMF measurements carried out in measuring cells containing static or stirred sample solutions. Recently, ion-selective electrode measurements performed in streaming solutions have become more and more important. In the following section, we will give a short survey, from a methodological point of view, on ion-selective electrode potentiometric measuring methods carried out in flowing solutions. A potentiometric measurement performed in a flowing solution may be only one stage of an ion-selective electrode analytical method. Furthermore, direct potentiometry (calibration procedures, addition and subtraction methods) and potentiometric titration can also be realized in flowing solutions. But there are special measuring techniques (injection method, triangle-programmed titration, etc.) which can only be applied in flowing solutions.

3. Analytical Methods Used in Flowing Solutions

a. General Considerations

The increasingly wide application of methods using ion-selective electrodes in flowing solutions can be explained by the following causes. First, from the point of detection with an ion-selective electrode, the measurement of EMF in flowing solutions has special advantages. The measuring surface of the electrode is continuously in contact with fresh reagent in flowing solutions. In this way, the electrode itself cannot exert any effect upon the composition of the solution to be measured, which may result in an increase in sensitivity. The moving solution ensures, in most cases, a small diffusion layer thickness, and thus a quicker electrode response can be attained. A significant advantage is that the reference electrode can be placed downstream of the indicator electrode, and leakage of ions from the reference electrode into the solution cannot disturb the detection. Second, certain types of measurements can only be carried out in flowing solutions, e.g., on-line methods employing ion-selective electrodes in monitoring and process control. Third, in the laboratory, serial analysis of a great number of samples can be simply solved by flow-through methods, thus avoiding the need to wash vessels and weigh samples.

A detector cell, suitable for measurements in flowing solutions, can be prepared in a simple way, since commercially available electrodes can be supplied with a flow-through cap. Some of the commonly used types of flow-through electrodes and potentiometric flow-through measuring cells are illustrated in Figure 6. The EMF measured in flow-through systems may be influenced[45] by the streaming potential (see Volume I, Chapter 3, I.L).

For measurements in flowing solutions, the methods of direct potentiometry are very suitable. By letting prepared standard solutions separately flow through the measuring cell, a calibration curve can be constructed. Then the sample solution, before and after premixing with the standard addition or subtraction solutions is let flow through the measuring cell; the unknown concentration can be calculated or be determined graphically in the way described earlier.

The standard addition measuring technique can be done directly in flowing solutions since addition can be performed continuously by introducing flow of standard solution into the flow of sample solution. The measurement is performed only after appropriate mixing. A disadvantage of this technique lies in the fact that it is sensitive to the flow rate. Multistandard addition can be performed in two ways, either with the help of standards of different concentration flowing at identical rates or with standards of identical concentration but different flow rates. If stabilization of the flow rates is ensured, the Gran linearization signal-processing method can be applied directly. Fleet

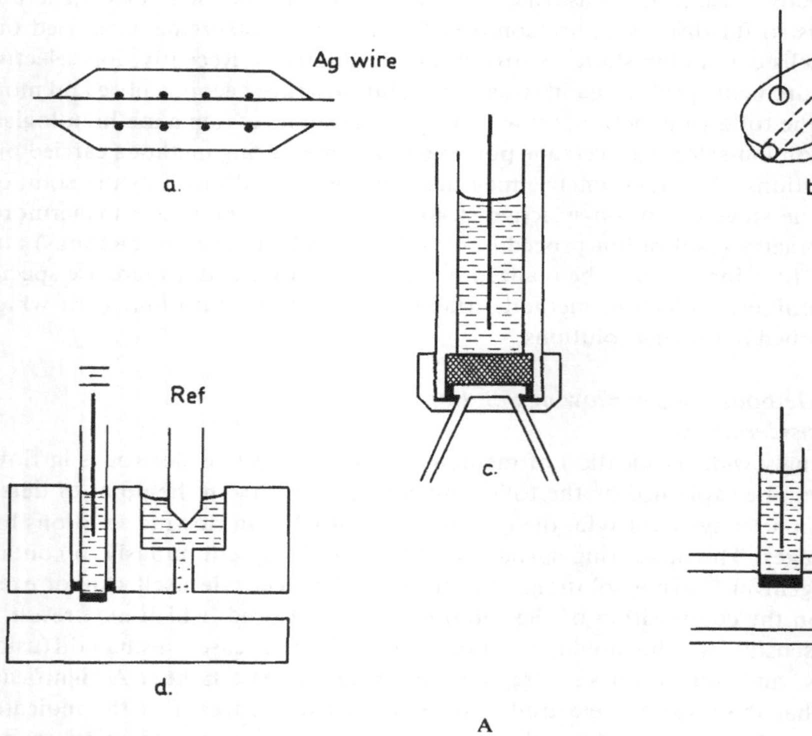

FIGURE 6. (a) Different types of flow-through potentiometric cells and electrodes: a, capillary glass electrode; b, capillary solid state electrode; c, flow-through cap electrode; d, e, flow-through cells. (From Fleet, B. and Ho, A. Y. W., in *Ion-selective Electrodes*, Pungor, N. E., Ed., Akademiai, Kiado, Budapest, 1973, 17. With permission.) (b) Flow-through electrode assembly: A, indicator electrode; B, reference electrode; C, solution ground; D, sensing membrane; E, Teflon® sleeve; F, plexiglass cap; G, washer. (From Llenado, R. A. and Rechnitz, G. A., *Anal. Chem.*, 45, 826, 1973. With permission.) (c) Flow-through measuring cell involving a microcapillary glass electrode and a Saturated Calomel Electrode: 1, salt bridge; 2, saturated KCl; 3, ceramic plug; 4, calomel; 5, mercury; 6, cable; 7, vacuum capillary; 8, rubber stopper; 9, plastic capillary; 10, solid state membrane; 11, inner reference solution.

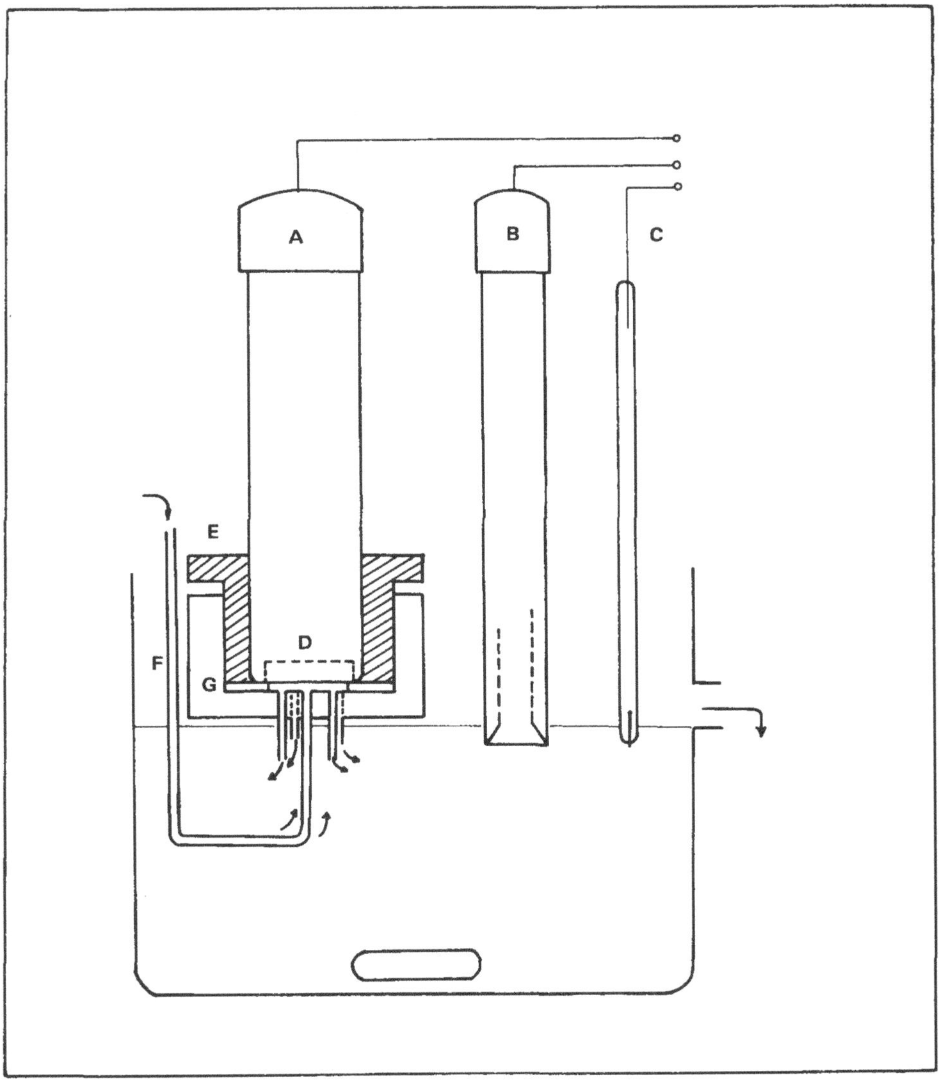

FIGURE 6 B.

and Ho[46] applied their semicontinuous concentration measuring method to an evaluation carried out with a computer, based on the Gran linearization technique. The addition is performed continuously; standard solutions of various concentrations are mixed at constant flow rate with the sample solution flowing also at a constant rate and supplied continuously with the appropriate background electrolyte.

For the semicontinuous analysis of flowing solutions, we also developed a standard addition and subtraction measuring technique.[47] The standard substance, to be added to solutions flowing with a constant rate and supplied with appropriate background solution, was prepared by continuous constant current electrolysis at various current densities i_1, i_2, ... In this way, we were able to solve the problem of keeping the flow rate of the standard solutions constant.

For continuous coulometric standard addition, the EMF (E) is given by the following formula, assuming that the current efficiency was 100% in the electrolysis:

$$E = E_o + S \log \frac{c_s v_s \pm k i_n}{v_s} \tag{11}$$

where v_s is the flow rate of the sample solution, c_s is the concentration of the primary ion in the sample solution, and i_n is the applied current density. S and E_o have their usual meanings; k is the constant characteristic of the electrolysis process and is related to the mass flow of the ion to be added, prepared at unit current density. In the case of subtraction, it contains also the stoichiometric factor of the species reducing the primary ion concentration.

The sign of product ki_n refers either to addition (+) or to subtraction (−).

From Equation 11, it is evident that

$$10^{E/S} = \text{constant} \left[\frac{c_s v_s \pm k i_n}{v_s} \right] \tag{12}$$

By applying different i_n values, we measure the EMF and plot $10^{E/S}$ as a function of i_n. From the i_n value obtained from the intercept of the line and knowledge of k and v_s, the concentration of the sample solution (c_s) can be calculated. The method of evaluation is explained by Figure 7.

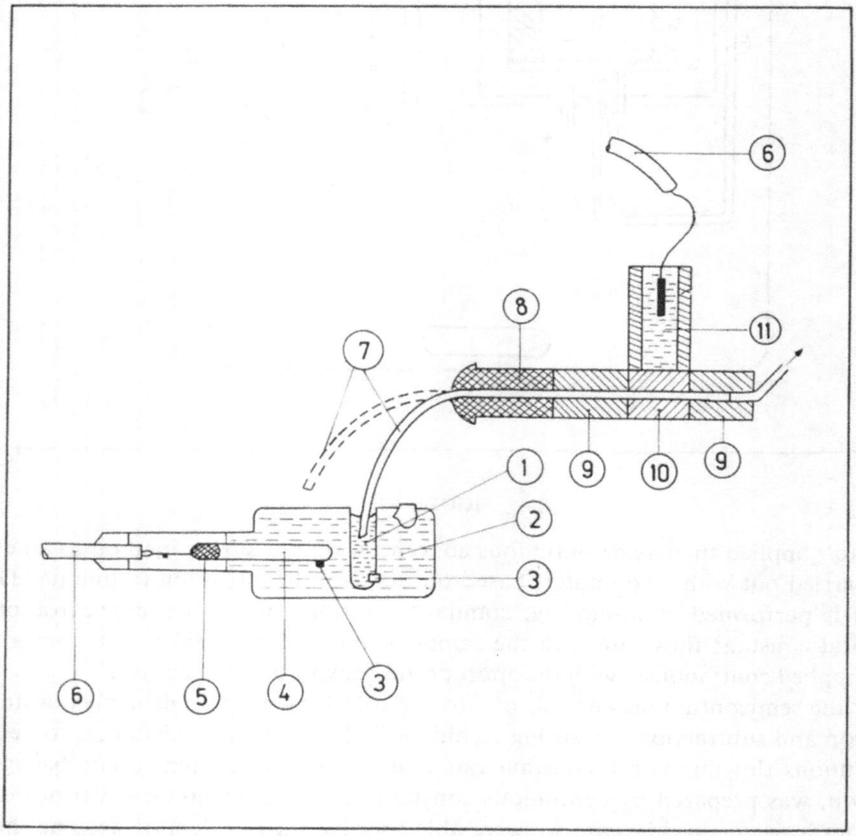

FIGURE 6 C.

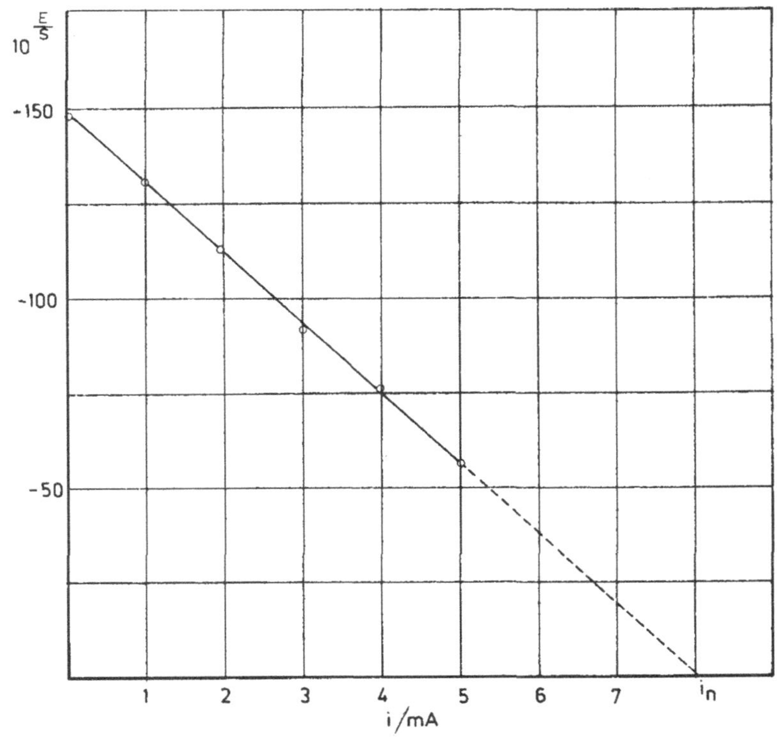

FIGURE 7. Gran plot diagram obtained in flowing solution with coulometric standard substraction method: V, 5 ml/min; ion generated, Ag^{+}; S, 58 mV/decade; flowing solution, 10^{-3} *M* KBr. (From Nagy, G., Feher, Z., Tóth, K., and Pungor, E., *Hung, Sci. Instrum.*, 41, 27, 1977. With permission.)

The method described was found to be suitable for the analysis of halide-containing samples carried out with standard subtraction and addition measuring techniques. The subtraction was realized by preparing silver ion reagent electrolytically. A brief description of the equipment used will be given later (see Section IV.A.3.c).

As mentioned previously, calibration, addition, and subtraction techniques can be applied without incurring greater difficulties using flow-through systems. The same is not true with potentiometric titration methods.

If the titration is carried out in a stirred solution at a constant rate of addition of reagent and detection is by a flow-through ion-selective electrode inserted into a recirculation by-pass system, then, besides the end-point indication problems arising from the dead-time of the by-pass circuit, one must contend with the loss of those advantages which are connected with the application of the flow-through system. Accordingly, it is evident that if we want to perform a titration in a flowing system in such a way that the advantages of the flowing system are maintained, then the reagent must be added to the sample solution, flowing with a constant rate, at a given point of the liquid-transporting tube section. In order to be able to carry out a complete titration in this way, it is important that the reagent flow should continuously increase from the beginning of the titration. Here, in contrast to classical titration methods, it is thus not the consumed total amount of the titrant which is decisive but the mass flow of the reagent in comparison with the mass flow of the titrand. Every different mass flow rate corresponds to a degree of titration. At the equivalence point, the mass flow of the sample is stoichiometrically equivalent to the mass flow of the reagent. The degree of titration can be increased by increasing the flow of the reagent, and reducing the flow leads to a reduction of the degree of titration. It is essential that the reaction upon which the titration is based should quantitatively take place in the space lying between the detector cell and the point where the reagent is added. In the interest of clear end-point detection, it is important also that in the direction of the flow, the mixing of the solutions should be kept to a minimum.

Since, due to the spatial separation, the detector follows the process with a certain delay, it is advisable to trace the total titration curves in flowing systems in such a way that we apply a mass flow of the reagent, which linearly increases with time, and record simultaneously against time the EMF from the ion-selective electrode detector cell.

Obtaining a continuously increasing mass flow of the reagent volumetrically can be realized either by increasing the volume flow rate or by linearly increasing the concentration of the reagent solution at the same flow rate.

To achieve the second of these procedures, a method was developed by Fleet and Ho[48] called "gradient titration," in which, under strict control of the ratio of flow rates, a reagent flow of constant volume rate with linearly increasing concentration could be ensured. Gradient titration belongs to a group of titration methods in which the determination is based on a predetermined EMF end point (titration to preset end point). It is evident that it is possible to trace the whole titration curve similarly by various burettes which can be programmed or through reagent generation by current programmed electrolysis, titration can be performed in a flow-through analysis channel.

We have tried to demonstrate briefly that many measuring techniques using ion-selective electrodes can be applied to flowing systems. However, the possibilities offered by flowing solutions make the development of further advantageous measuring techniques possible which would be impossible with static or stirred solution samples. Next, two of these measuring techniques will be described.

b. Triangle-programmed Titration Technique

The triangle-programmed titration method is the first of the two to be presented from the point of view of measuring technique.

Prior to the detailed discussion of the method, it is worth mentioning that evaluation based on the titration curve ensures a higher reliability than the preadjusted end-point titration. This is especially true if potentiometric indication is applied.

During titration of flowing solutions, the detector follows the progress of titration with a delay, owing to the spatial separation of detector and reagent addition point. Theoretically, this delay can be quite accurately known from knowledge of the reagent addition program, geometrical conditions of the appropriate cell parts, and the flow rate. This makes possible the accurate determination of the reagent mass flow associated with the equivalence point. The practical situation is, however, more complicated. The step-by-step increase of the reagent mass flow may be accompanied by the increase of the flow rate (e.g., volumetric-programmed reagent addition), and the "tailing" due to the reagent absorption on the wall of the tube also makes the accurate determination of the delay difficult. Significant uncertainty is caused in the determination of the reagent mass flow associated with the equivalence point of the program from the section between the starting point and the equivalence point. Much more disadvantageous is the determination of the reagent mass flow from the section between the equivalence point and the end of the titration (the instant of stopping the reagent addition), since the effect of stopping the reagent mass flow is not clearly visible on the signal/time curve of the detector (washing-off effect).

In order to eliminate the above-mentioned uncertainty factor, it is expedient to consider the special character of titrations performed in flowing solutions, namely, that

the degree of titration can be reduced by decreasing the ratio between the reagent and titrand mass flows. The triangle-programmed titration technique developed in our laboratory is based on this principle.[49-51] A rate/time program of reagent addition is applied, which results in two equivalence points on the signal/time recording of the detector. Accordingly, during the first period of the program, the system reaches an overtitration; then, by decreasing the reagent flow, the overtitration is reduced continuously to the equivalence point and further to the titration level at $t = 0$. Thus, a reagent flow/time program in the shape of an isosceles triangle, ensuring an overtitration, is employed. For the calculation of the reagent flow belonging to the equivalence point, two well-defined, well-identifiable points on the EMF vs. time graph, namely, the two equivalence points, are used. To establish the relation between the time of appearance of the equivalence points and the concentration of the flowing sample, the following facts should be considered. Let us suppose that a sample solution of constant concentration (c_s) during a titration is flowing through the analysis channel with a constant volume rate (v). In the course of the titration, the reagent flow (V_R) linearly increases up to time τ ($V_R = tn$), when n is constant. Then, for $t > \tau$, the reagent flow linearly decreases in the following way:

$$V_R = (2\tau - t)n \tag{13}$$

The complete flow-rate program is demonstrated for coulometric reagent generation in Figure 8. Suppose the following quantitative titration reaction takes place among the reagent (R) and sample (S):

$$aS + bR \longrightarrow dP_1 + gP_2 \tag{14}$$

where a,b,d, and g are stoichiometric numbers and P_1 and P_2 are products of reaction. At the equivalence point, an equality of mass flow exists, and

$$V_{RE} = c_s v \tag{15}$$

The time instant of its appearing can easily be obtained from the stoichiometry and the reagent mass flow program for a certain point of the analysis channel, for both equivalence points t_{E1} and t_{E2} under the following simplifying assumptions:

1. The mixing of the sample and the reagent is instantaneous and complete at the point of confluence.
2. The confluence of the two mass flows takes place in a section of infinitesimal thickness.
3. The chemical reaction takes place instantaneously and quantitatively, according to Equation 15.

Then,

$$t_{E_1} = \frac{a}{bn} \cdot c_s v + t_r \tag{16}$$

$$t_{E_2} = 2\tau - \frac{a}{bn} c_s v + t_r \tag{17}$$

where t_r is the delay time associated with a certain point of the channel ($t_r = W/v$, where W is the volume between the place of the confluence of the reagent and sample and the point of the channel in question).

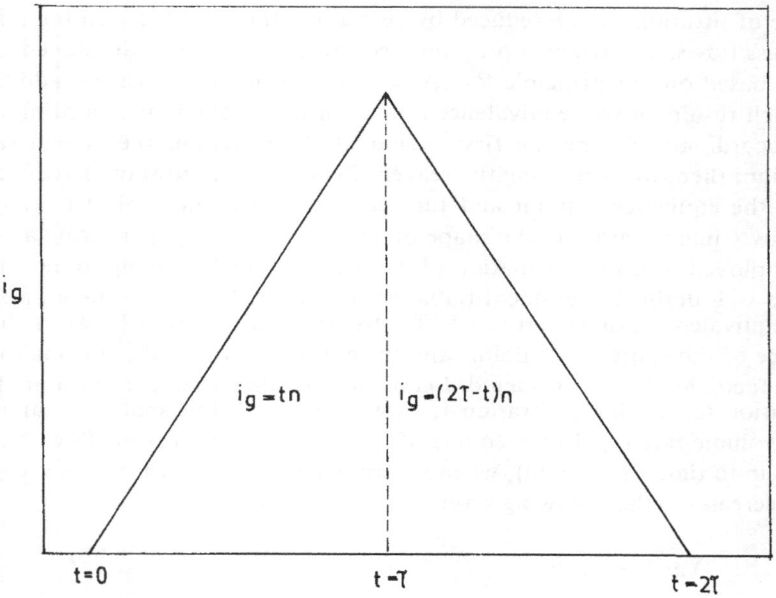

FIGURE 8. Isosceles triangle signal used for programming the reagent mass current. i_g = generating current, t = time, τ = time period while the reagent generation is increased, and n = a constant. (n = k di/dt where k represents the amount of reagent produced by a unit of charge.)

The time interval (Q) passed between the appearance of the two equivalence points is given by

$$Q = t_{E_2} - t_{E_1} = 2\tau - 2\frac{a}{bn}c_s v \qquad (18)$$

It can clearly be seen that between the Q value and the concentration of the flowing sample solution, a simple linear relationship exists. There is thus a possibility for accurate determination of concentration either by calculation or by a calibration curve recorded with standard solutions. In Figure 9, theoretical potentiometric titration curves (EMF vs. time) obtained by the triangle-programmed titration method are presented, as well as the method of their deduction for the case of a symmetrical titration curve. The various curves relate to different c_s concentrations. It can be seen that for a sample solution having c_4 concentration, even the highest ratio of the reagent/sample solution flow is not sufficient to reach the equivalence point. The curves also demonstrate clearly that a concentration variation causing relatively small change in EMF may produce a significant difference in the appropriate Q value when a proper reagent addition program is used. The triangle-programmed titration technique can be realized both with volumetric and coulometric reagent addition.

The argentimetric determinations carried out by current-programmed coulometric Ag^+ ion generation will be described in detail. The schematic design of the apparatus used is shown in Figure 10. The solution-carrying channel is made of different sections. The first part of the channel is an electrolysis cell serving for reagent generation. The two flow-through compartments of the electrolysis cell are separated by a cellophane membrane. Because of the direction of flow, it is obvious that the electrolysis products cannot mix. The other part of the analysis channel is the detector cell, which is separated from the generator cell by a drip vessel, ensuring the mixing of reagent and sample solutions and electrical separation of electrolysis cell and detector. The pro-

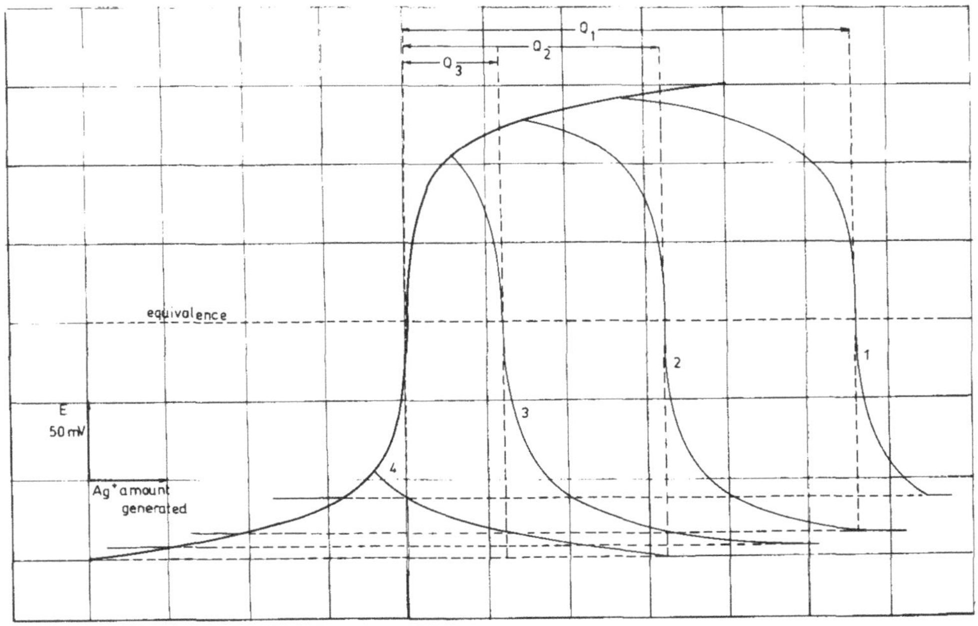

FIGURE 9. Theoretical programmed coulometric titration curves characteristic for potentiometric detection. (From Nagy, G., Feher, Z., Tóth, K., and Pungor, E., *Anal. Chim. Acta*, 91, 87, 1977. With permission.)

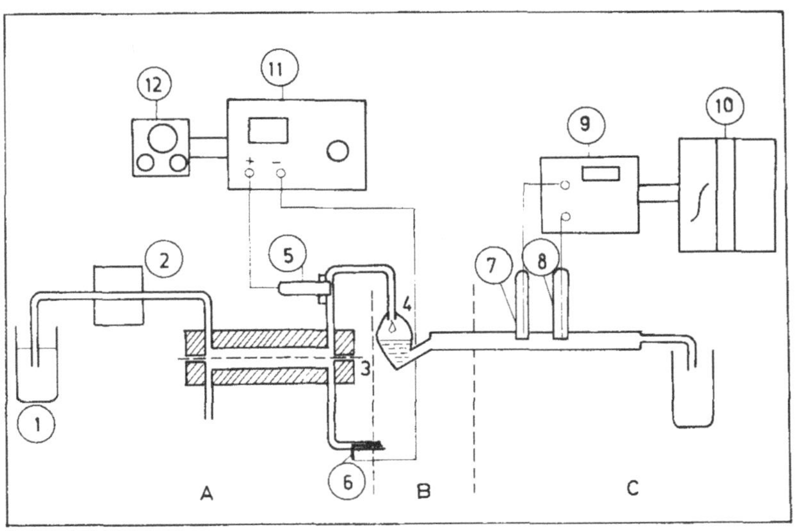

FIGURE 10. Block diagram of the experimental set-up constructed for programmed coulometric study: 1, solution reservoir; 2, peristaltic pump; 3, dialysis membrane; 4, drip vessel; 5, generating electrode; 6, counter electrode; 7, indicator electrode; 8, reference electrode; 9, measuring instrument; 10, recorder; 11, current generator; 12, signal generator. (From Nagy, G., Feher, Z., Tóth, K., and Pungor, E., *Anal. Chim. Acta*, 91, 97, 1977. With permission.)

grammed generating current is provided by instrumentation consisting of a function generator and current generator. The titration curve is continuously recorded by a pH meter and recorder.

For samples containing chloride ions, potentiometric titration curves are obtained (see Figure 11) if the sample solution supplied with the appropriate background electrolyte is passed through an analysis channel provided with an Ag anode-generating electrode. By employing Q values from the titration curves, a calibration curve can be obtained (Figure 12).

In special cases, this measuring technique also offers the possibility, by means of a suitable argentometric reagent generation program, for the simultaneous determination of various components reacting with silver ions (Figure 13). The applicability, accuracy, and reliability of the triangle-programmed titration technique can readily be appreciated.

c. Injection Technique

The second special measuring technique to be described here is also based on the principle of the flow-through channel and is called "the injection technique." At the end of the 1960s, this method was developed employing voltammetric detection.[52,53] This measuring technique, however, proved to be particularly advantageous with ion-selective detectors.[54-58] Recently, this method has been employed by several research groups for solving a number of different analytical tasks.[59-68]

In the injection technique, a solution of composition appropriate to the detector is flowed at constant volume flow rate in a flow-through analysis channel. Before the detector, at a special point, doses of solution of small volume or ions generated by

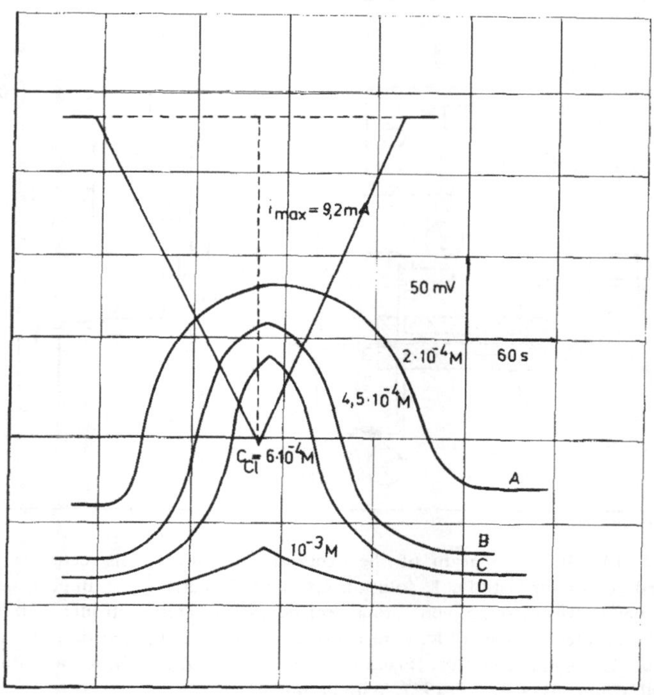

FIGURE 11. Argentometric programmed coulometric titration curves for chloride. Chloride concentration: A, 2.10^{-4} *M*; B, 4.5 10^{-4} *M*; C, 6.10^{-4} *M*; D, 10.10^{-4} *M*. (From Nagy, G., Feher, Z., Tóth, K., and Pungor, E., *Anal. Chim. Acta*, 91, 97, 1977. With permission.)

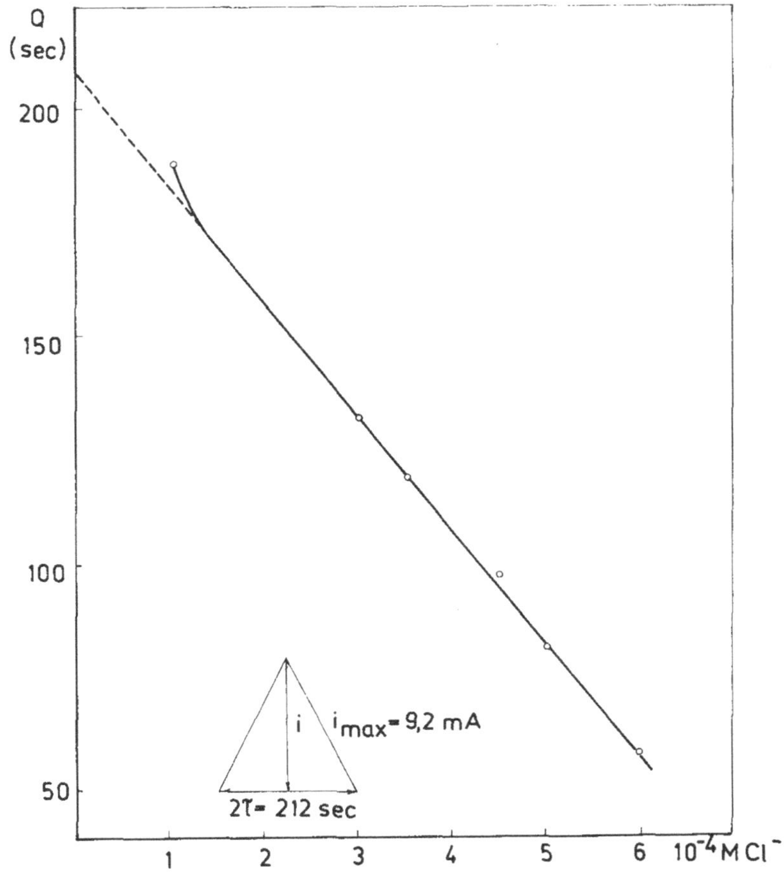

FIGURE 12. Calibration curve for the determination of chloride samples with programmed coulometric titration technique. Flow rate: 5.3 ml/min.

electrolysis are injected into the stream of solution. The peak-shaped transient signal-time curves which the detector senses correspond to the passing through the detector cell of the dose injected.

Chemical reactions between the components of the injected sample and the streaming solution or dilution effects due to the mixing have an influence on the profile of the concentration or ion activity in the solution flowing in the tube. Obviously, the transient signal appearing on the detector can only be used for analytical purposes if reproducible conditions and signals can be ensured by means of a special apparatus.

For reproducible conditions, the following are necessary:

1. The quantity of the sample injected should be strictly constant.
2. The flow rate of solution is similarly kept at a constant value with high accuracy.
3. The flowing conditions as well as the geometry of the tube section between the place of injection and detection should be chosen so that the individual doses flow through this section of tube in a reproducible manner.
4. Reproducible and efficient mixing conditions should be created.
5. Similar injection rates should be ensured.

The first condition can be provided with the required accuracy by means of additional units available (precision syringes, automatic burettes, and current-programmed coulometric units). With an appropriate peristaltic pump, the volume flow rate does not change significantly during a short period of time. The third condition is expected also to be fulfilled in the case of a simple flow-through system. However, the last two conditions may raise some problems. In our work, a separate mechanical mixing unit has been placed between the place of injection and detector cell, which ensures a complete and reproducible mixing and eliminates the alterations caused by the differences in injection rate. Ruzicka et al.[60,67] found another solution to this problem. They achieved complete mixing by properly choosing the flowing conditions, but for manual injections, the rate of injection must be reproducible at a constant value. In other work, reproducibility is ensured by using a semiautomatic injector.[65]

The schematic design of our measuring apparatus is shown in Figure 14. If into solution flowing at a constant rate and containing that ion in appropriate concentration to provide an electrode response a small quantity of the same ion is injected, then the EMF change time curve is peak shaped, owing to the change in ion activity. The transient signal obtained by injection technique is shown in Figure 15. The correlation between concentration and time is easy to describe after making a few assumptions.[53] The relation between EMF change (ΔE_t) and time can be given, according to the expected variation of ion activity following the injection, with the help of an equation describing the response of the ion-selective electrode, assuming a constant ionic

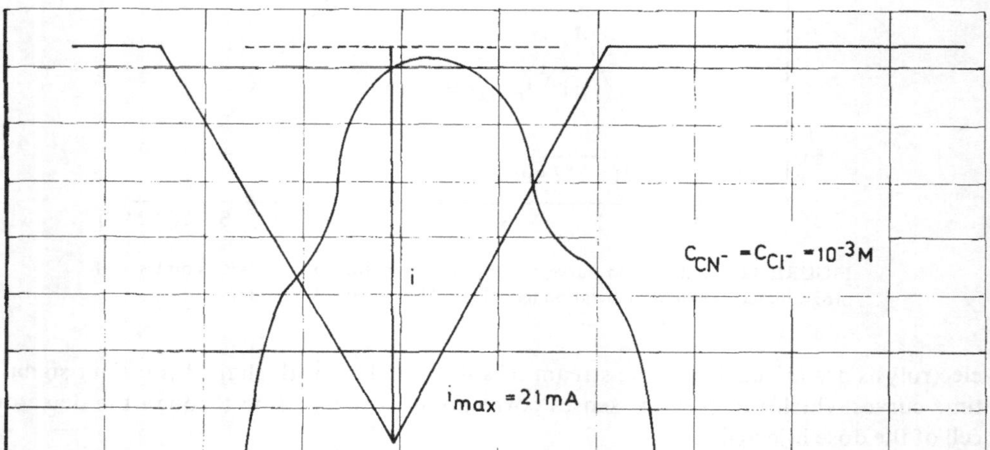

FIGURE 13. Programmed coulometric titration curve obtained in flowing solution containing cyanide and chloride ions. Flow rate: 3.9 ml/min. 2τ = 298 sec. (From Tóth, K., Nagy, G., Feher, Z., and Pungor, E., *Z. Anal. Chem.*, 282, 379, 1976. With permission.)

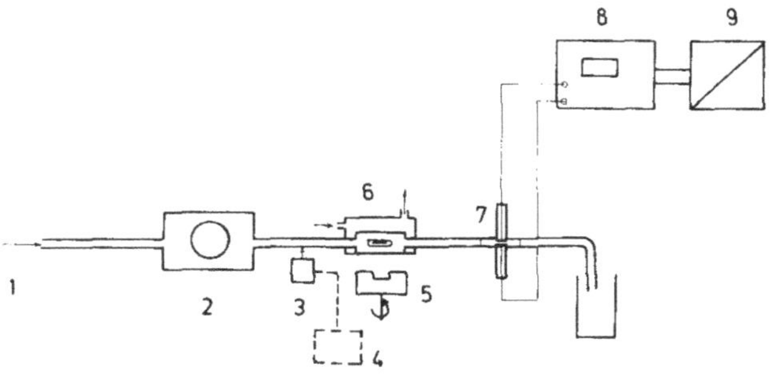

FIGURE 14. Block diagram of an experimental set-up constructed for the injection study: 1, solution reservoir; 2, peristaltic pump; 3, injection unit; 4, programmer; 5, stirrer, 6, thermostat jacket; 7, detector unit; 8, measuring instrument; 9, recorder. (From Tóth, K., Nagy, G., Feher, Z., and Pungor, E., *Z. Anal. Chem.*, 282, 379, 1976. With permission.)

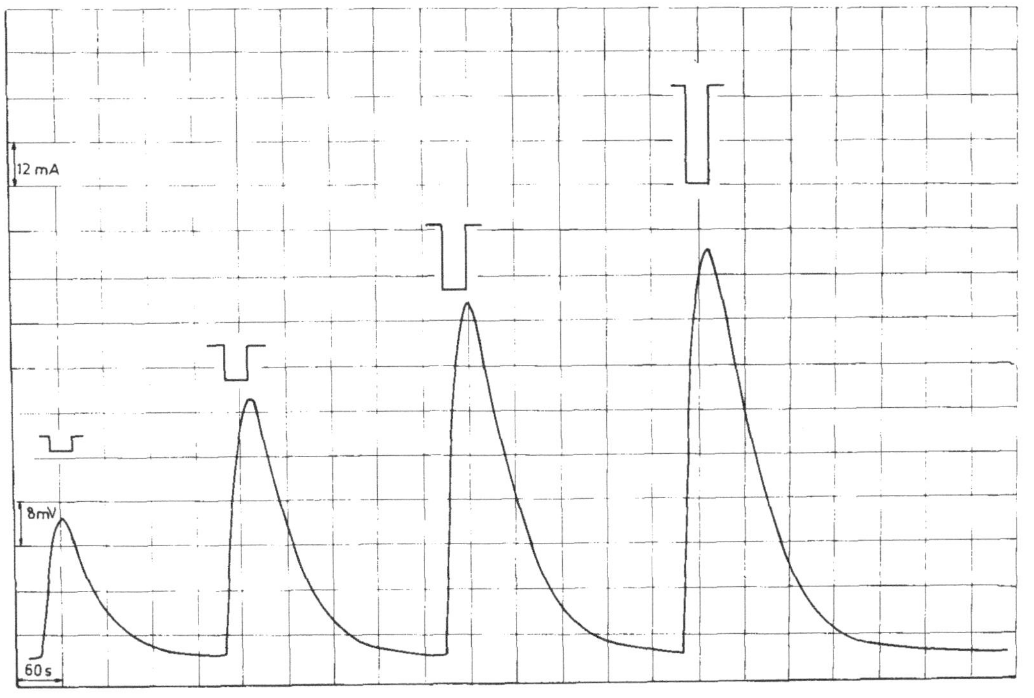

FIGURE 15. The signals of an ion-selective electrode recorded after coulometric injection. Flow rate: 5.3 ml/min. Flowing solution: 10^{-4} *M* $AgNO_3$ – 10^{-} *M* KNO_3 – 10^{-2} *M* HNO_3. Reagent injected: 1.26×10^{-6}, 2.52×10^{-6}, 5.40×10^{-6}, 8.40×10^{-6} mol Ag^+.

strength. Consequently, the concentration change (Δc_t) due to the injection at a time instant $t < \tau$

$$\Delta E_t = c_t - c_o = \frac{M}{v\tau} [1 - e^{-vt/W}] \tag{19}$$

while if $t \geqslant \tau$,

$$\Delta c_t = \frac{M}{v\tau}\,[1-e^{-v\tau/W}]\,e^{-v(t-\tau)/W} \tag{20}$$

where M is the moles of ions injected, v is the flow rate, W is the volume of mixer, and τ is a factor characteristic of the rate of injection, i.e., the interval between the time at which the sample solution containing the quantity of ions injected reaches the inlet of the mixer and the time at which the total dosage of sample solution passes over the inlet. Thus, introducing the factor representing the effect of the interfering ions on the operation of the ion-selective electrode if C_o is the primary ion concentration of the flowing solution, then

$$\Delta E_t = S \log \left[1 + \frac{\Delta c_t}{c_o + \sum_{j=i}^{n} k_{ij}\, c_j^{zi/zj}}\right] \tag{21}$$

The transient signal has a characteristic, easily evaluable feature, namely, the peak height at time $t = \tau$, ΔE_τ.

It is clear that both ΔE_t and ΔE_τ and other parameters of the transient signal (e.g., area under the peak) are dependent partly on the quantity of the primary ion injected and partly on the primary ion concentration of the flowing solution. Consequently, the transient signal is suitable for the determination of both of these important data in analysis. On the one hand, the transient signals can be used for the determination of ion concentration in a solution flowing with constant rate when an appropriate constant ion quantity is injected. On the other hand, based on the relationship between the ion quantity injected and the appropriate parameter of the transient signal, the serial analysis of individual sample solutions can be performed by employing streaming solutions of constant flow rate and suitable constant ion concentration.

The form of the transient signal becomes quite complicated if a chemical reaction takes place between the flowing solution and the sample injected, but it is still fundamentally affected by both the quantity of the sample injected and the concentration of the flowing solution. Consequently, by keeping the other parameters constant and by an appropriate choice of the experimental conditions, an extremely wide spectrum of analytical methods employing injection techniques can be developed. The determination of substances giving no electrode response, present either in the streaming solution or in the sample injected, is also possible.

The main advantage of the injection technique in comparison with other flow-through techniques based on signal formation at a steady state lies in its high specific speed of analysis. With steady state-type mechanized flow-through analyzers, the 40 to 60 samples-per-hour analysis rate can hardly be increased further even by the trick of sample separation by bubbles. The injection technique can easily achieve a speed of 200 to 360 samples per hour.[61] This is due to the fact that an addition of reagent needed to reach the steady state is not necessary. As soon as the detector signal decreases to its initial value, the apparatus is ready for the next sample to be injected.

A further advantage of the injection measuring technique is that the component injected is only in contact with the electrode for a short time; thus, the lifetime of the electrode can considerably be increased when samples damaging the electrode are to be measured. When a chemical reaction takes place between the flowing solution and the components of injected solutions, a substance contaminating the analysis channel or the electrode may be produced e.g., precipitate formation. In such a case, the superiority of the injection technique is indisputable. The flowing solution clears the channel between the injected doses of reagent.

The applicability of the method for serial analysis will be demonstrated by a practical example. The determination of samples containing cyanide ions at high concentration by means of a cyanide electrode must be carried out with diluted samples so that the strong cyanide solution does not destroy the electrode after a few analyses. Dilution, however, means a loss of time. This problem is solved by the application of injection technique so that dilution can be eliminated and a long lifetime of the electrode ensured. A suitable solution flowing through the analysis channel containing a cyanide ion-selective electrode has a concentration of 10^{-1} *M* of KNO_3 and 10^{-4} *M* of KI, with a pH value of 11. The solution should be free of oxygen. The analysis is performed by serial injection of cyanide sample solutions of 0.1-ml volume at a flow rate of 5 ml/min (v) and using a mixing chamber with a volume of 3 ml (W). Under these conditions, the determination can conveniently be carried out in the concentration range 10^{-1} to 10^{-2} *M*. The calibration curve of ΔE_T vs. log c_{CN} obtained with standard solutions is used for the evaluation.[55]

Table 2 lists some analytical problems solved by the injection technique developed by Pungor and co-workers.

Some measuring techniques employing ion-selective electrodes for primary ion determination have been discussed. These methods can be inserted in any other analytical process as constituent steps.

B. Methods with Previous Chemical Conversion of the Sample to a Species Detectable by the Indicator Electrode

In analytical practice, there are many methods based on ion-selective electrode detection which, as a first step in the analysis, require the determination of chemical conversion of the component. A short review of these techniques of chemical transformation will be presented, since the other steps of processes are similar to those discussed earlier.

A quantitative chemical conversion is necessitated under the following conditions:

1. There is no selective electrode available suitable for the direct detection of the component to be measured. Frequently, such a component is non-ionic; however, by a quantitative chemical reaction it can be transformed into ions detectable by ion-selective electrode.
2. The sample does not dissolve in solvents compatible with the ion-selective electrode.
3. The component can be converted by quantitative chemical reaction into a form which can more easily be measured with an ion-selective electrode.
4. Effects of components which disturb the operation of ion-selective electrodes can be decreased or eliminated by an appropriate chemical reaction.

In most cases, the chemical transformation is demanded by several of these requirements.

It must be noted that the first quantitative reaction of most analytical methods using ion-selective electrodes cannot be considered merely as a simple sample preparation step; therefore, they should be treated here. The reasons for this are as follows:

1. The conventional sample preparation steps (fusing, combustion, and aqueous digestion) can be combined with ion-selective electrode detection, usually with some variation.
2. The development of the chemical conversion is due to the introduction of ion-selective electrode detection.

TABLE 2

A Survey on the Application of the Injection Technique with Potentiometric Detection

Flowing solution	Solution injected	Component determined in	Aim of the determination	Detector electrode
$KI + KNO_3$	I^-	Injected solution	I^- in pharmaceuticals	I^-
$KI + KNO_3$	$I^- + Cl^-$	Injected solution	I^- in Cl^-	I^-
$KI + KNO_3$	CN^-	Injected solution	CN^- samples	CN^-
$KCl + KNO_3$	Chloropromazine HCl, diethazine HCl, melipramine HCl	Injected solution	Cl^- content in pharmaceuticals	Cl^-
$KBr + KNO_3$	Gastrixone, methylhomatropinium bromide	Injected solution	Br^- in pharmaceuticals	Br^-
TRIS® (pH = 7.0)	Urea	Injected solution	Urea samples	Urea electrode
TRIS + urea	Urease enzyme	Injected solution	Urease enzyme activity	NH_4^+
Thio-cholin-ester	Cholinesterase enzyme	Injected solution	Cholinesterase enzyme activity	S^{2-}
$Cl^- + KNO_3$	Cl^-	Flowing solution	Cl^- content of drinking water	Cl^-
$CN^- + KNO_3$	CN^-	Flowing solution	CN^- content of sewage	CN^-
I^- + buffer	Glucose	Flowing solution	Glucose samples	Glucose enzyme electrode
$CN^- + KNO_3$	Ag^+	Flowing solution	CN^- content of sewage	CN^-, Ag^+
S^{2-} + NaOH	Ag^+	Flowing solution	S^{2-} content of samples	S^{2-}
S^{2-} + NaOH	Hg^{2+}	Flowing solution	S^{2-} content of samples	S^{2-}

Adapted from Pungor, E., Tóth, K., Nagy, G., and Feher, Z., in *Ion-selective electrodes*, Pungor, E., Ed., Akademiai Kiado, Budapest, 1977.

3. The choice of the conditions for applying the reaction is influenced to a large degree by the requirements of ion-selective electrode detection.

Organic molecules containing nondissociable halogens can be measured after decomposition by halide determination with ion-selective electrodes. Thus, methods using ion-selective electrodes combined with combustion methods used in elementary analysis, e.g., the oxygen flask method,[19,69] are widely applicable in the analysis of pesticides and pharmaceuticals containing halogens.

It is most likely that the various combustion and digesting methods used in classical analysis (aqueous decomposition, oxidizing decomposition in the Parr bomb, and catalytic decomposition carried out in an oxygen stream) could be successfully adapted for detection with ion-selective electrodes if quantitative chemical transformation and absorption can be ensured, although this has been proved only in a few cases.

From the point of view of ion-selective electrode detection, proper selection of the absorption medium is an important question. In several cases studied, an absorption medium different from that used in classical analysis has been suggested, and the reagents following the absorption step should be chosen so that they are compatible with the detector. As an example, a modified analytical method applying Schoniger combustion will be described which is convenient both for practical analysis and students' laboratory experiments.

According to the method, 15 to 20 mg of a solid, pulverized halogen-containing compound is weighed out. The substance is introduced onto a previously ignited platinum grid of a Schoniger flask of 1 to 1.5 l packed in ashless filter paper in an L shape. In the case of hardly burnable samples, it is suggested that the sample be mixed with saccharose in the ratio 1:1. Previously, the flask should be prepared by adding 10 ml distilled water and 5 ml ammonia solution (1:1) and filling with oxygen gas. Then the paper pack containing the sample is lit and the sample burned. After cooling, the combustion gases are absorbed by shaking. The solution in the flask is weakly acidified with 1:1 nitric acid; then, once the carbon dioxide has been evolved, the halide content of the sample can be determined by any of the previously detailed measuring techniques using the appropriate ion-selective electrode. According to our experiments, the sulfuric or acetic acid neutralization suggested in the standard Schoniger method[70] is not suitable, especially when samples contain chloride ion.

Hassan[71] developed a method for the determination of the chlorine and bromine content of organic compounds in which halogen is liberated by combustion in a flask filled with oxygen. As absorption agent, alkaline hydrogen peroxide is used, and following absorption and neutralization, the ion concentration is measured by potentiometric titration in a 1:1 mixture of dioxan-water.

If the purpose of the burning in oxygen is the detection of fluoride ion,[72] after absorption and before acidification, the contents of the flask should be poured into a plastic beaker. The acidification can be carried out with hydrochloric acid. The determination can be made with a fluoride ion-selective electrode, after applying TISAB solution by titration with lanthanum nitrate or by employing one of the techniques of direct potentiometry.

Methods employing ion-selective electrodes which include combustion in a continuous stream of oxygen have also been developed. Potman and Dahmen[73] have described an apparatus and method suitable for the determination of the chloride and bromide content of small quantities of volatile organic samples. The samples are injected into the apparatus by syringe. After decomposition in an oxygen stream at 1000°C in a platinum and quartz vessel, the combustion gases are carried over by the oxygen stream into an absorbent solution containing acetic acid (80%), hydrogen peroxide (1.2%), and nitric acid, plus a trace of mercury (II) chloride or bromide. Then the determination of bromide and chloride is performed by titrating with Hg^{2+} solution. The end-point detection is done with the Ag_2S-based ion-selective sulfide electrode. Employing the method of titration to a preset end point, the apparatus and method are suitable for serial analysis of a great number of samples without renewing the absorbent solution.

Some analytical methods based on ion-selective electrodes employ steps such as conventional combustion in a bomb or digestion by alkali fusion. A few examples connected with the determination of fluorine content will be given for demonstration.

For the determination of the fluorine content of carbon samples, a Parr combustion bomb and alkali-fusion method have been described by Thomas and co-workers.[74] In the first case, the procedure is the following: 1 g of the well-pulverized carbon sample (100 mesh) is mixed with 0.25 g benzoic acid in a fused quartz sample holder, and this is placed in a Parr combustion bomb containing 5 ml of 1 *M* sodium hydroxide. The samples are burnt at oxygen pressure of 28 atm. Following the combustion, the contents of the bomb are washed into a plastic vessel, the pH adjusted to 5.0 to 5.2 with 0.25 *M* sulfuric acid, and the carbon dioxide outgased by heating. Analysis is carried out with the addition of TISAB solution at pH 6 by means of the standard addition technique.

In the alkali-fusion method, 3 g of sample is mixed with 5 g of water-free crystalline sodium carbonate. The mixture is placed in a platinum crucible and covered with 2 g of sodium carbonate, then heated at 475°C for 24 hr. The contents of the crucible are melted at 1000°C. After cooling and extraction with warm water at a pH 5.0 to 5.2, the CO_2 is expelled, and the F^- determination is performed with a fluoride ion-selective electrode, using TISAB solution and the standard addition technique.

Baker found that alkali fusion was applicable to plant samples. The standard addition method is used for measuring the fluoride content of previously dried plant samples fused with sodium hydroxide.[75]

The Kjeldahl digestion, originally developed for the determination of protein content of beer in 1883, can be applied also in the preparational stages of methods using ion-selective electrodes. However, the Kjeldahl method using acid-base titration can hardly be included with methods operating with ion-selective electrodes. Since the appearance of the ion-selective ammonium and ammonia gas electrodes, the Kjeldahl digestion has gained importance. With the application of the above electrodes, the time-consuming distillation separation step can be eliminated. The Kjeldahl digestion can be performed continuously in flowing systems;[76] thus, the special advantage of detection with the ammonia-sensitive gas electrode gains greater significance. Buckee[77] succeeded in attaining an analysis rate of 60 samples per hour.

For the Kjeldahl digestion, solutions of high-ionic strength are necessary. If a conventional ammonia gas electrode is used for a long time in flowing solution as the detector in analytical processes employing Kjeldahl digestion, then, due to the different osmotic conditions existing on the two sides of the membrane, a so-called "osmotic drift" may appear, originating from the water transport. To reduce this effect, one can alternatively (a) dilute the sample solutions or (b) increase the ionic strength of the inner buffer solution used in the ammonia gas electrode by adding an inert electrolyte such as sodium sulfate. The latter alternative is preferable, since dilution could result in a sample solution having an ammonia content below the detection limit of the electrode.

Catalysts used in the digestion which would interfere with the measurement must be masked. For example, ammonia complexes formed with copper ion catalyst and mercury (II) ions are brought into forms with iodide ions or EDTA, respectively, which do not interfere with the detection.

Special cases of chemical conversion are those methods in which the ion species to be measured is brought into a form which can be measured in a more advantageous way by means of a selective sensor. A very simple example is the measurement of the sulfite content of samples after acidification, with a sulfur dioxide electrode. The same can be said of the measurement of carbonate ions with CO_2 electrode or the concentration of ammonium ions by an ammonia gas electrode.

In certain cases, redox reactions can also be applied as transformation steps. Two examples will be given. As mentioned previously, the nitrate-selective electrodes available are not of high selectivity. Chloride, bicarbonate, and iodide ions, which are often present in the sample, interfere with the determination of nitrate. Therefore, in many cases, it is advisable to perform the measurement with an analytical procedure which employs a reduction step and ammonia gas electrode as detector. The reduction step can be performed in several ways. Mertens et al.[78] found Dewarda's alloy bound to PVC or polystyrene as matrix to be very suitable for this purpose. It is advisable to perform the reduction at pH 13. The method can be applied to flowing systems by using the flow scheme demonstrated in Figure 16. Using this scheme, 20 samples could be analyzed in 1 hr. Naturally, the lifetime of the reductor column is of critical importance. The authors were successful in preparing a column which ensured a 5-day lifetime with a 100% reduction under the conditions of analysis.

In the second example, we shall show that the determination of two ionic species in the presence of each other can be solved with the help of a method using ion-selective indicator electrodes, even when no electrodes of entirely adequate selectivity are at our disposal for the measurement of either ion species. Morie et al.[79] devised a method for the determination of nitrate and nitrite in the presence of each other. The sample solution is acidified with phosphoric acid, and the potential of the ion-selective nitrate

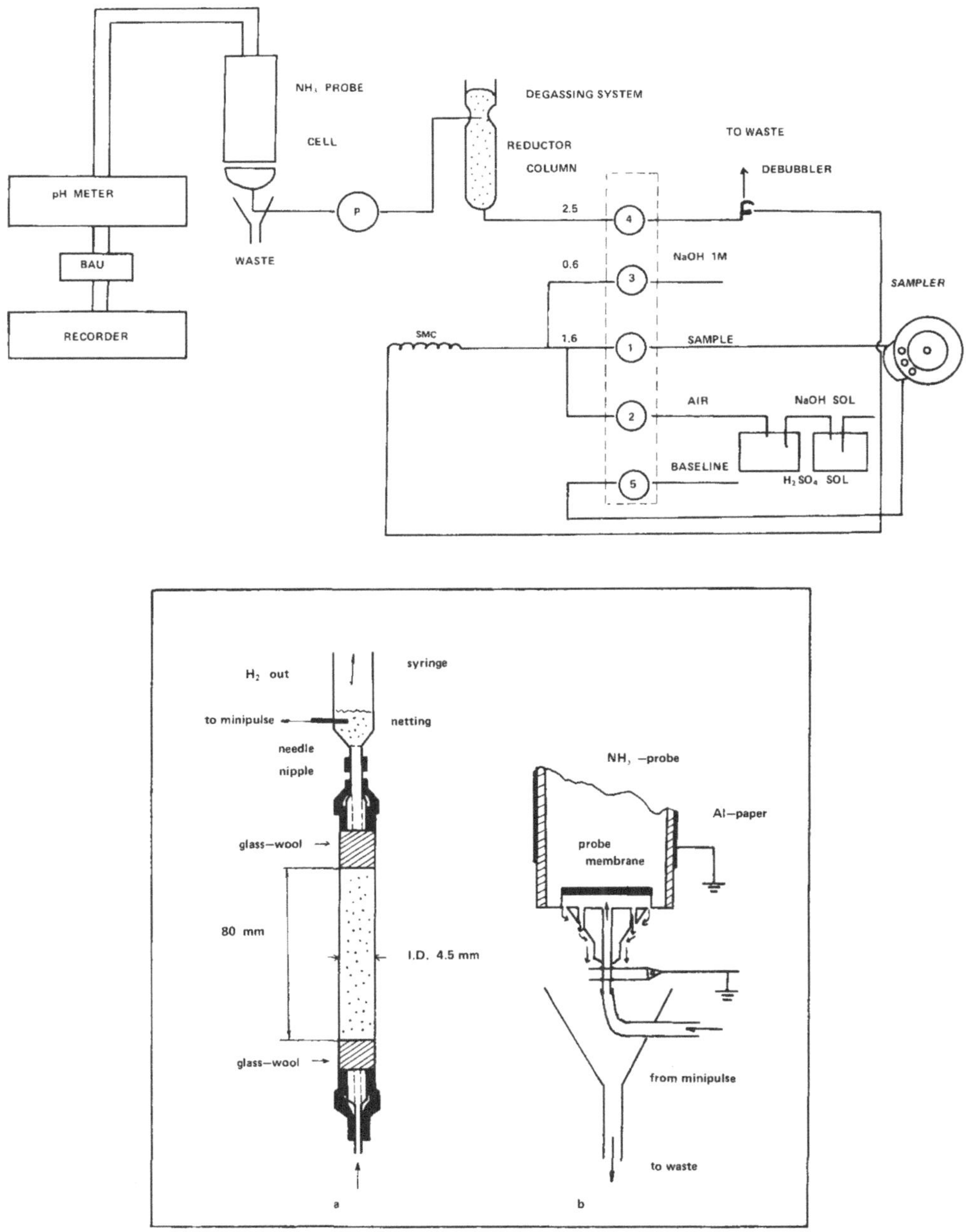

FIGURE 16. Flow diagram of the automatic system for nitrate determination: a, the reduction column with the degassing system; b, the open flow system. BAU = base-line adjustment unit. P = minipulse pump with adjustable speed. (From Mertens, J., Van den Winkel, P., and Massart, D. L., *Anal. Chem.*, 47, 522, 1975. With permission.)

electrode is measured, which is defined by the activity of both the nitrate and nitrite ions present in the sample (E). Thereafter, the nitrite ions present in the solution are oxidized with potassium permanganate solution and from the measurement of the EMF (E_t), the total nitrate concentration $[NO_3^-]_t$ is determined. The concentration of the nitrate ions $[NO_3^-]$ originally present in the sample can be determined from

$$[NO_3^-] = 10^{(E_t - E)/S} - k \frac{V_2}{V_1} \frac{(NO_3^-)\,t}{1 - k} \qquad (22)$$

Where V_1 is the volume of the sample solution, V_2 is the solution volume in the second EMF measurement, k is the selectivity coefficient of nitrate over nitrite, and S is the slope. Thus, with the knowledge of the total nitrate concentration, the original nitrite concentration can also be obtained. The accuracy of the method fundamentally depends on the accuracy with which the selectivity coefficient can be obtained under the given conditions. Only with careful operation and over a narrow concentration range can an accurate result be expected like that obtained by the authors (standard deviation 1.6% NO_3^-, 6.1% NO_2^-).

A special group of chemical conversions are enzyme-catalyzed reactions:

$$S + R \xrightarrow{\textit{enzyme}} P_1 + P_2 \qquad (23)$$

During the reaction, the activity of a component to be detected by the ion-selective electrode changes in the reaction mixture. Only those processes in which one of the reaction products or the substrate itself is detected can be classed among those analytical methods involving direct chemical transformation. Those analytical methods in which the reagent taking part in the enzyme-catalyzed process is detected are included in the group of indirect analytical methods.

By the help of enzyme-catalyzed reactions, measuring methods of very high selectivity can be developed for the quantitative determination of organic components. Both the advantages and drawbacks of this kind of analytical method derive from the character of enzyme reactions:

1. A great advantage is the high selectivity originating from the substrate and reaction specificity of the enzyme catalyst.
2. Ion-selective electrode detection gives the advantage that, in most cases, the presence of proteins in solution does not cause any interference; thus, in general, no separation is required.
3. One of the drawbacks is, however, that usually a long incubation time is necessary for the completion of the reaction.
4. The applicability of analytical methods using enzyme-catalyzed reactions is limited by the relatively high cost, scarcity, instability, and unreliability of the enzymes.

The significance of the last factor, however, is gradually diminishing, owing to progress in fine chemical production and to the use of immobilized enzymes. Presumably, the number of of ion-selective electrode methods using enzyme catalysts will rise as a result of the development of new enzymes.

The analytical processes devised to date have been based mainly on hydrolysis and oxidation-reduction (decarboxylation) reactions. Carbon dioxide, ammonia gas, or ammonium and cyanide ion-selective sensors suitable to the nature of the reaction have been used for detection.

The most widely used and studied analytical methods are those employing the enzyme urease and various ammonia gas or ammonium ion-selective sensors for urea measurement.

In the analytical methods[80,81] developed for the determination of urea in the 1960s, cation-selective glass electrodes were used to measure the concentration of NH_4^+ ions, the product of enzyme-catalyzed reaction:

$$NH_2 - \overset{\overset{\displaystyle O}{\|}}{C} - NH_2 + 2H_2O \xrightarrow{urease} CO_3^{2-} + 2NH_4^{\;+} \qquad (24)$$

Recently, the method has been improved, and ammonia gas and ammonium ion-selective electrodes of higher selectivity have been used as detectors. In the opinion of Rogers and Pool,[82] the application of the ammonia gas-sensing electrode is more advantageous, since its operation is not influenced by K^+, Na^+, and Ag^+ ions, which interfere with the function of cation-selective glass electrodes; consequently, they can be applied in media of various compositions.

The NH_3-sensitive gas electrode was used[82] for the determination of urea-nitrogen in sewage-water samples when ammonium ions were also present. The NH_3 content of the sample solution was examined prior to the enzymatic hydrolysis. The pretreatment of the urea sample consisted of incubation with urease enzyme solution at a constant temperature (25°C), in the presence of TRIS buffer of 0.1 *M* ionic strength for a given period of time. The measurement with the gas electrode was done at pH 12.

A potentiometric method for the detection of urea content in blood and blood serum was devised by Hansen and Ruzicka[83] with the application of the air-gap ammonia gas electrode. For calibration, a urea standard or sample (200 μl) treated with 1 ml TRIS buffer solution containing 7.4 units of urease activity was incubated for 2 min at 26°C. The mixture was transferred to the electrode vessel, and 1 ml 0.2 *M* sodium hydroxide was added. The EMF of the cell was measured with continuous stirring. The standard deviation was found to be under 2.4%. Because of the design of the air-gap electrode neither the protein content of the samples nor their differing alkali metal ion concentrations caused problems during the measurement.

The ion-selective electrode methods applying enzyme-catalyzed reactions can also be utilized in flowing systems. However, the measurement is mostly performed without complete conversion, except for very rapid reactions, and the change in ion activity or concentration during a certain period of time is detected. According to the classification adopted in this chapter, these methods are classed among processes based on reaction rates (kinetic method). Some other enzyme-catalyzed reactions playing a role in ion-selective electrode measurements are dealt with later.

V. INDIRECT ANALYTICAL METHODS

The favorable features of measurements with ion-selective electrodes initiated the development of indirect methods for the determination of materials giving no well-defined electrode response. The material to be determined is involved in a quantitative chemical reaction. The change of the concentration or activity of a reagent or an indicating material as a result of the reaction is followed by an ion-selective electrode, and the concentration of the sample is determined by calculation or calibration curve, based on the amount of reagent added and the stoichiometry of the chemical reaction.

As has already been noted, it is difficult to separate sharply indirect measuring methods employing ion-selective electrodes and methods measuring the primary ion. It is also difficult to distinguish between primary ion measuring methods combined with chemical transformation and indirect measurements with ion-selective electrodes. It is worth mentioning that the difference lies in the fact that in the case of the primary ion measurements, the sample itself is transformed into a component detectable by an ion-selective electrode; by indirect methods, a change in the activity of another ion is detected by the ion-selective electrode which by the quantitative reaction is related to the concentration of the sample.

Obviously, methods by which the ion-selective electrode detects species other than the material to be determined or a component obtained from it through chemical transformation may have a great many variations. However, the measuring methods using ion-selective electrodes developed for special purposes are sometimes less favorable than the classical methods. Consequently, only a few indirect analytical methods developed following the appearance of a new ion-selective electrode have found widespread application

Table 3 lists some indirect methods for the analysis of components which cannot be carried out directly by the ion-selective electrodes available to date. The methods listed mainly use titration techniques.

It can be seen from Table 3 that in indirect methods, besides the exceptionally favorable ion-selective electrodes (fluoride, iodide), the copper and lead electrodes are often applied. The copper electrode can advantageously be used as indicator electrode in determinations involving complexing agents, while the ion-selective lead electrode has a role in indirect methods accompanied by precipitate formation.

In most simple, indirect methods, the procedure is as follows: The reagent detectable with the ion-selective electrode and reacting with the component to be determined is added directly to the sample solution, and the EMF of the measuring cell containing the ion-selective electrode is measured.

Direct calibration measuring techniques and addition, subtraction, and titration techniques can be applied both in stationary and flowing solutions. In most cases, the concentration of sample is calculated by considering the stoichiometry of the analytical reaction or, occasionally, by direct calibration. However, the indirect method employing ion-selective electrodes is frequently more complicated. For example, the method developed by Selig and co-workers[84] for the determination of arsenate ion employs a back titration technique. Lanthanum nitrate is added to the sample in excess, and the excess is determined by measuring the EMF with an ion-selective fluoride electrode, using a reagent addition procedure with sodium fluoride. A wide spectrum of analytical methods similar to the above can be developed; among these, a method for phosphate may be mentioned.

An interesting technique for aluminum determination is described by Homolo and James,[21] which is suitable for the determination of aluminum content of white water in a paper factory. A given volume of properly treated sample solution is introduced into a measuring cell containing an ion-selective fluoride electrode. Fluoride solution of given concentration is added to the sample solution to bring the EMF to a certain constant value ascribable to a given free fluoride ion activity. The aluminum ion concentration of the sample solution is determined from a calibration curve obtained with standard aluminum samples. Accurate results can be obtained by this method only if in the examined aluminum concentration range, the stoichiometry of the complex $AlF_x^{(3-x)+}$ produced by the fluoride addition remains identical for different samples. The accuracy of results obtained seems to verify that under conditions used (pH = 4 to 5.5), the same complex composition is ascribable to a constant free fluoride ion activity. From seven measurements at an Al^{3+} concentration of 10 mg/l, the standard deviation of the results was 1.1%.

The interfering effects of Ca^{2+}, Fe^{3+}, and Mg^{2+} ions present, due to complexation and precipitate formation, were eliminated by adding an excess of these ions to both the sample and calibrating solutions.

A considerable number of analytical methods based on the detection of reaction partner with ion-selective electrodes developed so far are based on detection with ion-selective iodide electrodes and on the reaction of the iodide-iodine redox system. This cannot be accidental. The iodine-iodide redox system — owing to its good reversibility and favorable redox potential — is capable of reacting quantitatively with the majority of both oxidizing and reducing materials. A wide variety of different classical and

TABLE 3

Characteristic Data of Some Indirect Titrimetric Analytical Methods Applying Ion-selective Electrodes

Component determined	Type of electrode	Type of titration	Media, supporting material	Titrant	Lower limit of determination (*M*)
Aluminum	Fluoride	Complexometric	Acetate buffer	NaF	10^{-3}
Barium	Copper	Chelatometric	Cu CDTA	CDTA	10^{-4}
Calcium (in the presence of Mg)	Copper	Chelatometric	Cu EGTA	EGTA	10^{-4}
Chelate-forming	Copper	Chelatometric		$Cu(NO_3)_2$	
Cobalt (II)	Copper	Chelatometric	Cu EDTA	EDTA	
Lithium	Fluoride	Precipitation	Alcohol	NH_4F	10^{-1}
Magnesium	Copper	Chelatometric	Cu EDTA	EDTA	10^{-4}
Manganese (II)	Copper	Chelatometric	Cu EDTA	EDTA	
Mercaptans, other organic compounds	Sulfide, Ag_2S	Precipitation		$AgNO_3$	
Molybdate	Lead	Precipitation		$Pb(ClO_4)_2$	
Nickel (II)	Copper	Chelatometric	Cu TEPA	TEPA	10^{-4}
Phosphate	Fluoride	Precipitation	$La(NO_3)_3$, acetate buffer	NaF	5×10^{-3}
Rare earth metals	Fluoride	Precipitation		NaF	10^{-4}
Selenide	Lead		SAOB	$Pb(ClO_4)_2$	10^{-4}
Strontium	Copper	Chelatometric	Cu EDTA	EDTA	5×10^{-3}
Sulfate	Lead	Precipitation	1:1 Methanol—water (pH = 4—5)	$Pb(ClO_4)_2$	5×10^{-3}
Telluride	Lead		SAOB	$Pb(ClO_4)_2$	10^{-4}
Tungstates	Lead	Precipitation		$Pb(ClO_4)_2$	
Vanadium (II)	Copper	Chelatometric	Cu EDTA	EDTA	10^{-4}
Zinc	Copper	Chelatometric	Cu TEPA	TEPA	10^{-4}
Zirconate	Fluoride	Precipitation		NaF	

Note: CDTA = 1,2-cyclohexylene diamine tetra acetic acid. TEPA = Tetraethylene pentamine. EGTA = Ethylene-glycol-bis(2-amino-ethyl ether).

Adapted from *Orion Research Inc., Analytical Methods Guide*, 5th ed., Cambridge, Mass., February 1973.

instrumental analytical methods is used in practice. From the point of view of ion-selective detection, the iodine-iodide redox system has a further advantage, in that it can be detected by an electrode having very favorable electrochemical properties, namely, the precipitate-based iodide electrode. Accordingly, simple, selective potentiometric measuring techniques can be worked out for the detection of a number of materials, so that in the first step, the material in question is reacted with the reagent containing the oxidized or reduced form of the redox system; the change in iodide ion activity occurring during the reaction is measured by an iodide electrode. When measuring reducing substances, the process is as follows: The sample solution is reacted with iodine solution under appropriate conditions, and the activity of iodide ions produced during the reaction is determined. Thus, the concentration of the reducing component can be determined indirectly. The sensitivity of the method depends on how far iodide-free iodine solution can be produced. A method based on the above description has been developed by Christova et al.[85] for measuring arsenite, sulfite, ascorbic acid, hydrazine, and hydroxylamine. The iodide-free iodine solution needed for the determination was produced by flowing the reagent solution through an ion-exchanger resin (Dowex® 1-X8 [chloride form]) previously saturated with ethanolic iodine solution. The determination was carried out in a monochloracetate buffer at pH 2, so the interference of iodine hydrolysis was eliminated. The possibility of measuring oxidizing materials is given from the decrease of iodide ion activity as a result of the following reaction:[86]

$$3\,I^- + \text{oxidizing component} \longrightarrow I_3^- + \text{product} \qquad (25)$$

The conditions of reaction are different from those of classical iodometric determinations, for an iodide ion excess greater by orders of magnitude than the sample concentration cannot be applied, since then the decrease cannot be detected by an ion-selective electrode with appropriate accuracy. A special measuring technique for the detection of the oxidizing component is described below.

In the course of investigating the structure of carbohydrates, primary information is provided by the amount of periodic acid consumption. Honda et al.[87] worked out a method for the measurement of periodic acid consumption by direct potentiometry with an ion-selective electrode. In carrying out the method, 1.0 ml solution of the carbohydrate (5×10^{-4} *M*) is reacted with an excess of sodium metaperiodate reagent solution (1 ml of 5×10^{-3} *M*, pH 6) containing 0.2 *M* acetate buffer. After keeping the reaction mixture at 25°C for 1 hr, it is reacted with an excess of potassium iodide solution (5×10^{-3} *M*) at pH 6.

$$IO_4^- + 2I^- + H_2O = IO_3^- + I_2 + 2OH^- \qquad (26)$$

After the above reaction takes place (5 min), the iodine is extracted with carbon tetrachloride; the iodide ion concentration is measured with the iodide electrode by direct potentiometry; the periodic acid consumption can then be calculated. By this method, it is possible to analyze samples of 1.0 to 10^{-2} μmol.

Indirect analytical methods often can be used also in functional group analysis of organic compounds. Hassan[88] has described a method employing an ion-selective chloride electrode as detector for the determination of amino groups. The organic sample containing the amino group is dissolved in anhydrous ether, and dry HCl gas is bubbled through the sample, which converts the amino group to its chloride salt. After evaporation of the excess of HCl gas by heating the prepared sample *in vacuo*, the chloride content of the dry chloride salts can be determined by argentimetric titration in 50% dioxan-water, with the chloride electrode for end-point detection. The results obtained for some primary amine compounds employing the micro-scale method described are collected in Table 4.

A rarely used analytical method employing an ion-selective electrode is described by Hadjiioannou and co-workers.[89] The method, which uses the ion-selective electrode as an end-point-indicating sensor, belongs to the group of catalytic titrations. It was developed to measure the concentration of EDTA in dilute solutions (7.5×10^{-6} to 10^{-3} *M*), but the analysis of various metal ions can be performed using this technique. As indicator reaction, the reduction of periodate ions with diethylamine is employed. The periodate activity is followed with an ion-selective perchlorate electrode. The EDTA operates as an inhibitor in the periodate reaction, while manganese (II) ions catalyze the reaction. If the partners of indicator reaction are led into the sample solution containing the EDTA, the periodate activity does not change (or only to a small degree, owing to the EDTA being present in excess). When manganese (II) solution is used as titrant, it reacts instantaneously with the EDTA, and the reaction which results in a decrease of periodate activity starts to take place when the titrant reaches an excess. The sudden change in the potential of the ion-selective electrode indicates the equivalence point. In fact, the end-point indication is performed with the ion-selective electrode measuring the activity of a material which itself does not take part in the titration

TABLE 4

Potentiometric Microdetermination of the Primary Amino Group in Some Organic Compounds Using Chloride Ion-selective Electrodes

Compound	Amino group (%) Calculated	Amino group (%) Found	Recovery (%)
p-Aminobenzoic acid	11.68	11.8	101.0
		11.9	101.9
p-Aminoacetophenone	11.85	11.7	98.7
		11.7	98.7
p-Aminobenzene	8.12	8.0	98.5
		8.2	101.0
p-Aminophenol	14.68	14.4	98.1
		14.5	98.8
o-Aminophenol	14.68	14.7	100.1
		14.5	98.8
Anthranilic acid	11.68	11.7	100.2
		11.7	100.2
p-Chloraniline	12.55	12.5	99.6
		12.3	98.0
α-Naphylamine	11.19	11.5	102.8
		11.2	100.1
m-Nitroaniline	11.59	11.6	100.1
		11.5	99.2
Benzidine	17.39	16.9	97.2
		17.1	98.3
o-Phenylenediamine	29.63	29.3	98.9
		29.3	98.9

Adapted from Hassan, S. S. M., Z. *Anal Chem.*, 270, 1974.

reaction. The end-point indication is quite precise, owing to the catalyzing effect. In all probability, the exceptionally high accuracy in the low concentration range (10^{-6} to 10^{-4} M— 1 to 2%), which cannot be attained by complexometric titrations with color indicator, is similarly due to the catalyzing effect.

VI. RATE METHODS EMPLOYING ION-SELECTIVE ELECTRODES

Methods based on reaction kinetic parameters or rate methods, being based on the relation between the rate of reaction and the concentration of different substances of the solution, are capable of solving a wider spectrum of analytical problems than methods based on the concentration signal measured under stationary conditions. Rate methods can be used to measure the activity of catalysts or the concentration of individual components present even only in trace amounts, which affect the operation of the catalyst. In spite of this, methods based on reaction kinetic parameters constitute only a relatively small proportion of those employed in analytical practice, and the proportion of these employing ion-selective electrodes is low.

Generally speaking the low popularity of analytical methods based on reaction kinetic parameters can be related to the following facts:

1. The rate of reaction may be influenced also by many components present in the sample. Thus, an analytical method based on the measurement of the concentra-

tion signal dependent directly on the concentration of the component to be determined is preferable.

2. Temperature, pH, and other parameters often have dramatic effects on the reaction rate; therefore, these parameters must be kept at constant values.
3. An easily applicable analytical method based on reaction kinetic parameters can be worked out only for reactions where the reaction rate lies in a certain range. Very fast reaction rates can be followed only with special equipment. When using an ion-selective electrode as detector, the response time of the electrode would be a limiting feature without a reaction-quenching step. Moreover, for slow reactions, the considerable time consumption of methods based on reaction kinetic parameters weakens their competitiveness.
4. The rather sophisticated laboratory instruments needed for this kind of investigation were devised later than those required for other methods; thus, it is easy to understand why the number of methods based on reaction kinetic parameters is rather low.
5. The kinetics of most reactions is less known in general than their stoichiometry. The relationship between reaction rate and concentration is often complicated.

A. Rate Methods with Transient Signal Evaluation

In one form of ion-selective electrode detection, the evaluation of the results is done from a transient signal. The chemical reaction takes place in the potentiometric measuring cell, and, simultaneously, the change of the EMF in time is continuously followed. The ion-selective electrode gives a potential signal, depending either on the increase in the activity of ions being formed during the reaction or on the activity decrease of a component. According to which of these possibilities occurs in the course of the reaction, the shape of the EMF vs. time function differs. The evaluation of results is made from the transient signal characteristic of the rate of reaction at a certain instant. In general, it is expedient to use the dE/dt value extrapolated to zero time. The concentration or activity of a catalyst, for example, can be determined with the help of a calibration graph, in which the dE/dt value extrapolated to its starting value ($t = 0$) is plotted against the concentration or activity of appropriate standard solutions.

Naturally, the shape of the EMF vs. time function observed during the reaction is dependent upon the kinetics of the reaction, on the reaction order, and on whether the potential of the indicator electrode is influenced by the activity of a component produced or used up in the chemical reaction. The logarithmic relationship existing between the indicating electrode potential and the ion activity influences the shape of the transient signal in a way which will now be considered.

If, with strictly first-order kinetics, the reaction rate constant does not depend on the initial concentration (c_o) of the ion to be measured with the ion-selective electrode, then the slope of the transient curve dE/dt vs. t is constant and is not suitable for the determination of concentration c_o. This means that

$$\frac{dE}{dt} = - S k \qquad (27)$$

where S is the slope of the calibration curve of the electrode and k is the rate constant. Thus, determination of the concentration of a species taking part in the reaction can only be done with ion-selective electrodes from the dE/dt value taken at $t = 0$ if the concentration of the component to be determined influences the rate constant. In any other case, linearization of the signal must be done previous to solving the analytical problem from the dE/dt readings.

Special care must be taken in the adaptation of rate methods originally worked out for linear signal-transforming detectors (e.g., voltammetric methods) to ion-selective electrode detection. As an example, a rate method which uses an ion-selective electrode detector and employs transient signal evaluation is that used to follow the rate of reaction

$$H_2O_2 + 3I^- + 2H^+ \longrightarrow 2H_2O + I_3^- \quad (28)$$

With an iodide electrode, it is possible to determine the concentration of components or the activity of catalyst influencing the reaction rate.

Altinata and Pekin[90] have developed a very sensitive rate method for molybdenum and tungsten determination. A given volume of sample, containing also sulfuric acid and potassium iodide in appropriate amounts, is placed into the measuring cell containing an iodide-selective electrode. A certain quantity of hydrogen peroxide is added to the sample, and the transient signal (EMF vs. time) is recorded continuously; the initial slope (dE/dt) at $t = 0$ is obtained. Owing to the sensitivity of the detection with the ion-selective electrode and the catalytic effect, both ions can be determined at concentrations as low as 0.004 μg/ml.

Radic[91] has worked out a rate method for the determination of aluminum-containing samples of small volume (2 ml) and low concentration (10^{-8} to 10^{-6} *M*), in which the ion-selective fluoride electrode is used for detection. Initially, it is immersed in a solution composed of 2 ml sodium fluoride solution (10^{-2} or 10^{-3} *M*), 2 ml acetate buffer of pH 5, and 25 ml distilled water. Then, 2 ml of sample is added to the potentiometric measuring cell, and the EMF value, determined by the fluoride ion activity, is recorded continuously. The evaluation of the results from a calibration graph, which is made by plotting the initial slope of the transient signal recorded in a standard solution against concentration values of the individual standards, allows the aluminum concentration to be read off. The calibration graph was found to be linear in the range of 8 to 300 nmol aluminum.

B. Rate Methods with Difference Signal Evaluation

The second group of rate methods involves analytical methods in which the evaluation is done on the basis of a difference signal. The relation between the change in reaction-rate-dependent signal produced in a given time interval and the concentration of the appropriate component of the sample or catalyst activity is determined, with identical reaction times for both standards and samples. It is advisable to carry out the reaction in the potentiometric measuring cell itself so that during the progress of the reaction, a well-measurable potential difference is established. With this procedure, the disturbing effect of the changes due to adding the solution to the cell and the displacement of the electrode is avoided. It is advisable to select the reaction time so that the progress of the reaction results in a potential change of 10 to 50 mV. The optimum reaction time may depend greatly on the properties of the reaction as well as on the electrochemical behavior of the detecting electrode. When the reaction time is long, separate chambers should be used for the reaction and for the potential measurement. Accordingly, sample solutions should be allowed in the measuring cell only for the time of detection; for relatively rapid reactions, the exact reaction time is adjusted so that the reaction is quenched by adding a suitable reagent at the appropriate instant.

Difference signal based analytical methods became especially important with the widespread use of the flow-through analysis channel, in which exact adjustment of reaction time can easily be made and the conditions necessary for the difference signal evaluation can be ensured. Furthermore, the time of analysis can be significantly reduced.

The reaction times obtainable with flow-through analysis channels limit the applicability of methods developed on the above principle. Thus, for example very rapid reactions, an infinitesimal reaction zone, a very high flow rate, and a detector cell of extremely small dead volume would be necessary, while for slow reactions, a very low flow rate, adjustment of which is rather tedious, and a very long reaction zone are required. In flow-through systems, analytical methods based on reaction rates that are catalyzed by an enzyme are very often applied. The reason for this lies in the fact that the rates of numerous enzyme-catalyzed processes are favorable as regards the parameters of the widely used flow-through systems. For enzyme-catalyzed reactions, ion-selective electrode detection gives special advantages, in contrast to optical detection methods in which the protein component of the solution often causes turbidity and hence light absorption. Several analytical methods will now be described in more detail. Rechnitz and Papastathopoulos[92] developed an interesting reaction-rate difference method for the measurement of the total cholesterol concentration of clinical serum samples. The procedure is a good example of the advantageous combination of the flow-through channel principle, detection by means of ion-selective electrodes, and the reaction-rate method. Since the new method demonstrates clearly the possibilities for achieving significant favorable conditions in chemical analysis, a detailed description of this new technique will be presented.

The evaluation of the method is carried out on the basis of the difference signal of the EMF, whose magnitude depends on the rate of the following reaction sequence:

$$\text{Cholesterol esters} + H_2O \xrightarrow[\text{ester hydrolase}]{\text{cholesterol}} \text{Free cholesterol} + \text{fatty acid} \tag{29}$$

$$\text{Free cholesterol} + O_2 \xrightarrow[\text{oxidase}]{\text{cholesterol}} \text{cholest-4-en-3-one} + H_2O_2 \tag{30}$$

$$H_2O_2 + 2I^- + 2H^+ \xrightarrow{\text{Mo(VI)}} I_2 + H_2O \tag{31}$$

The measurement is performed in a Technicon AutoAnalyzer® (Technicon Instr., Tarrytown, N.Y.), incorporating as detector a flow-through ion-selective iodide electrode. The flow diagram of the technique is shown in Figure 17. There are two reaction zones. The two enzyme-catalyzed reactions[29,30] taking place in the first step are quenched by acid added at the point (a). Thereafter, at point (b), there begins the reaction leading to the reduction of the iodide concentration, i.e., the electrode-responding reaction.[31]

A calibration curve is determined in which the change in the electrode potential, occurring as a result of the reduction of iodide ion activity, is plotted as a function of the total cholesterol concentration of the standard solutions. Using this method, a rate of 20 samples per hour can be achieved. The accuracy which may be atained with clinical samples is remarkable, even though, due to the instability of parameters such as flow rate, temperature, and enzyme activity, frequent recalibration is necessary. Quenching of the reaction by acidifying the solution until the proteins precipitate does not interfere with the operation of the ion-selective electrode. Only after examination of about 40 to 50 samples must the system be cleaned from the components precipitated by an alkali wash. With optical detection, this method of quenching the reaction would probably lead to a serious source of error.

For the measurement of urea concentration, a reaction-rate method with similar difference signal evaluation can be applied.

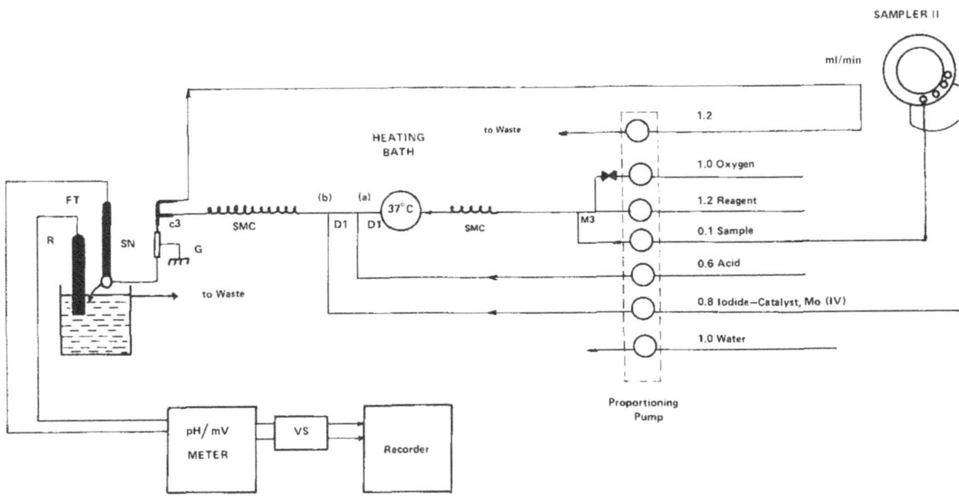

FIGURE 17. Schematic diagram of automated analysis system used for cholesterol determination: D1, fittings; SMC, mixing coil; c3, debubbler; SN, stainless steel ground contact; FT, sensing electrode; R, reference electrode; VS, voltage suppressor. (From Papastathopoulos, D. S. and Rechnitz, G. A., *Anal. Chem.*, 47, 1792, 1975. With permission.)

The reaction

$$\text{Urea} + 2H_2O \xrightarrow{\text{urease}} CO_3^{2-} + 2NH_4^+ \tag{32}$$

established in the Technicon Auto Analyser is quenched with concentrated alkali solution, thereby shifting the equilibrium $NH_4^+ \rightleftharpoons NH_3$ in the direction of free ammonia. This allows an ammonia gas electrode of favorable selectivity to be incorporated as detector into the equipment.

The schematic design of the analyzer is shown in Figure 18. The method was developed by Llenado and Rechnitz[93] and can be applied with success to the analysis of blood serum samples. With suitable alterations, the above method can be applied to various reactions. The general diagram of apparatus which can be developed in this way and suitable for the measurement of enzyme activity is shown in Figure 19.

Llenado and Rechnitz[94] used the following reaction for measurement of the activity of glucosidase enzyme:

$$\text{Amygdalin} + H_2O \xrightarrow{\text{glucoidase}} \text{Benzaldehyde} + 2\,\text{glucose} + HCN \tag{33}$$

with the cyanide-selective electrode as detector. As quenching reagent, 0.25 M NaOH was used, which simultaneously adjusts the pH to the value necessary for the cyanide measurement.

The version of the method developed for the measurement of rhodanese is based upon the reaction

$$S_2O_3^{2-} + CN^- \xrightarrow{\text{rhodanese}} SCN^- + SO_3^{2-} \tag{34}$$

Measurement of glucose oxidase activity is made on the basis of the two-step reaction

$$\beta\text{-D-glucose} + H_2O + O_2 \xrightarrow[\text{oxidase}]{\text{glucose}} \text{D-gluconic acid} + H_2O_2 \tag{35}$$

$$H_2O_2 + 2H^+ + 2I^- \xrightarrow{Mo(VI)} 2\,H_2O + I_2 \quad (36)$$

$$\text{Adenosine} \xrightarrow[\text{deaminase}]{\text{adenosine}} \text{Inosine} + NH_3 \quad (37)$$

Quenching of the reaction was performed by the addition of 0.1 *M* perchloric acid. The calibration curve is prepared on the basis of the electrode potential changes (ΔE), which are due to the changes in ion activity occurring in the course of the reaction. The equipment continuously traces the EMF developed in the measuring cell during the measurement. After the passage of one sample, a stationary value can be observed. If the sample is changed for another too quickly, then this stationary value cannot totally be attained, and the accuracy of the method may decrease.

Rechnitz and co-workers[95] used the following reaction for determination of the activity of the adenosine deaminase enzyme:

This determination was also carried out in the Technicon AutoAnalyzer with an ammonia electrode as detector. Measurement of the enzyme activity is based on the relation between the potential change of the ammonia electrode corresponding to one enzyme sample and the enzyme activity. For quenching the reaction and adjustment of the pH necessary for the detection, concentrated NaOH is added as a continuous stream at the end of the reactor tube, thus ensuring a 13-min incubation time.

It is evident that it is not always necessary to quench the reaction, even with relatively rapid reactions[86,96,97] carried out in flowing systems. The time elapsed between the beginning of the reaction and the measurement of the potential is defined by the flow rate and the volume of the reaction zone. The solution for quenching the reaction is mainly necessary to ensure a pH value suitable for the detection. The use of an enzyme-inactivating solution may, in certain cases, decrease the danger of contaminating the detector electrode.

At first glance, it seems advantageous to use two identical ion-selective electrodes incorporated into the flowing system. One electrode, the so-called "reference electrode," is inserted into the zone before the beginning of the reaction, while the other, the indicator electrode, is incorporated after the reaction zone. It is evident that if, during the reaction, a change occurs in the activity of the ion species detected by the electrodes, then the potential difference between the two electrodes would depend on the rate of the reaction. This arrangement of electrodes eliminates the measuring error caused by the changes in the diffusion potential inherent in a conventional reference electrode.

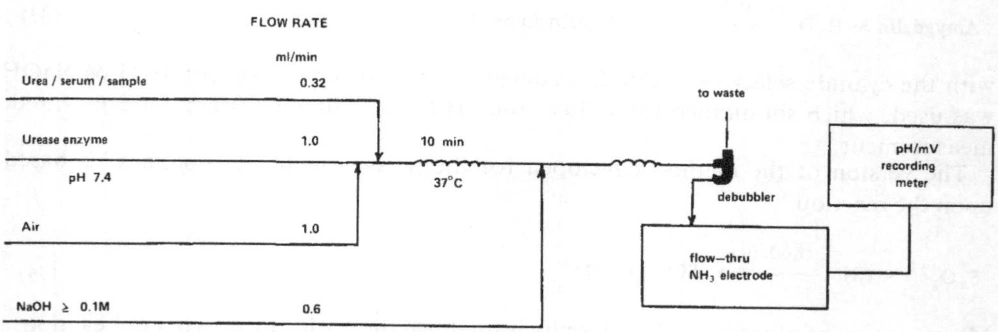

FIGURE 18. Schematic diagram of the automated continuous flow analysis system for urea. (From Llenado, R. A. and Rechnitz, G. A., *Anal. Chem.*, 46, 1109, 1974. With permission.)

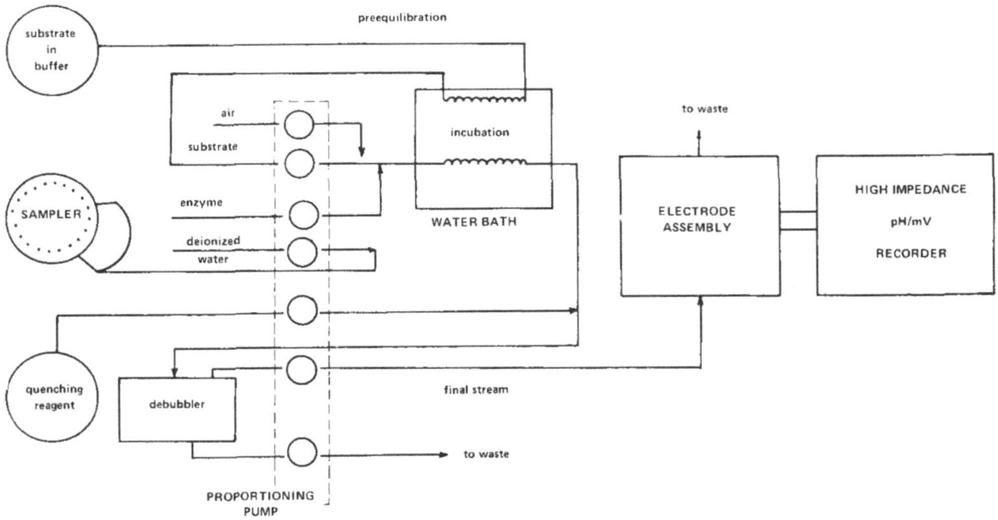

FIGURE 19. Schematic diagram of the generally used continuous flow apparatus for enzyme activity measurements. (From Llenado, R. A. and Rechnitz, G. A., *Anal. Chem.*, 45, 826, 1973. With permission.)

However, the interfering effect of species responding to the electrode and present in various concentrations cannot be eliminated by the above arrangement, since this decreases the slope of the calibration curve of the ion-selective electrode. This is easily understood by considering the following. Let a_i be the activity of the ion which defines the potential of the ion-selective electrode in the solution contacting the reference electrode (situated before the reaction zone), and Δa_i the activity change which takes place in the course of the reaction. The effect of the interfering ions is taken into consideration by using the usual k_{ij} term. Thus, the change in the electrode potential (ΔE_R) due to the changes in the activity during the reaction can be written as follows:

$$\begin{aligned} \Delta E_R = {} & (E_{o1} + S \log [a_i + \Delta a_i + \Sigma k_{ij} a_j^{zi/zj}] \\ & - E_{o2} + S' \log [a_i + \Sigma k_{ij} a_j^{zi/zj}]) \\ & - (E_{o1} + S \log [a_i + \Sigma k_{ij} a_j^{zi/zj}] \\ & - E_{o2} + S' \log [a_i + \Sigma k_{ij} a_j^{zi/zj}]) \end{aligned} \tag{38}$$

If S= S, therefore

$$\Delta E_R = S \log \frac{(a_i + \Delta a_i + \Sigma k_{ij} a_j^{zi/zj})}{(a_i + \Sigma k_{ij} a_j^{zi/zj})} \tag{39}$$

$$\Delta E_R = (E_{o1} + S\log[a_i + \Delta a_i + \Sigma k_{ij}\, a_j^{zi/zj}] - E_{ref}) -$$

$$(E_{o1} + S\log[a_i + \Sigma k_{ij}\, a_j^{zi/zj}] - E_{ref})$$

$$= S\log\frac{(a_i + \Delta a_i + \Sigma k_{ij}\, a_j^{zi/zj})}{(a_i + \Sigma k_{ij}\, a_j^{zi/zj})} \qquad (40)$$

Thus, identical expressions[39,40] are obtained for both electrode arrangements. The application of an identical ion-selective electrode as a reference electrode cannot therefore eliminate the disturbing medium effect. Accordingly, it should be considered whether the above-mentioned advantage is so significant that it is worth the complication of involving measurement of the potential difference between two high-resistance electrodes, requiring the application of a more complicated measuring device than the generally used pH meter.

The principle of the flow-through analytical channel combined with detection by ion-selective electrodes may simplify methods based on reaction rate. It can be expected that similar and new methods will be used widely in analytical practice in the future for the determination of the activity of enzymes and catalysts and for the measurement of components which exert an effect upon the catalyst function.

REFERENCES

1. Pungor, E. and Toth, K., *Anal. Chim. Acta,* 47, 291, 1969.
2. Pungor, E., Toth, K., Nagy, G., and Feher, Z., in *Ion-selective Electrodes,* Pungor, E., Ed., Akademiai Kiado, Budapest, 1977, 67.
3. Llenado, R. A. and Rechnitz, G. A., *Anal. Chem.,* 43, 1457, 1971.
4. Ruzicka, J. and Hansen, E. H., *Anal. Chim. Acta,* 78, 145, 1975.
5. Ruzicka, J., Hansen, E. H., and Zagatto, E. A., *Anal. Chim. Acta,* 88, 1, 1977.
6. Cheng, K. L. and Cheng, K., *Mikrochimica Acta,* 385, 1974.
7. Baumann, E. W., *Anal. Chem.,* 48, 548, 1976.
8. Hulanicki, A., Lawandowski, R., and Lewenstam, A., *Analyst,* 101, 939, 1976.
9. Hulanicki, A., Lawandowski, R., and Maj. M., *Anal. Chim. Acta,* 69, 409, 1974.
10. Guilbault, G. G. and Hrabankova, E., *Anal. Chim. Acta,* 52, 287, 1970.
11. Mascini, M., *Analyst,* 98, 325, 1973.
12. Milham, P. J., Award, A. S., Paull, R. E., and Bull, J. H., *Analyst,* 95, 751, 1970.
13. Guilbault, G. G., Nagy, G., and Kuan, S. S., *Anal. Chim. Acta,* 67, 195, 1973.
14. Fiedler, U., Hansen, E., and Ruzicka, J., *Anal. Chim. Acta,* 74, 423, 1975.
15. Hallsworth, A. S., Weatherell, J. A., and Dentsch, D., *Anal. Chem.,* 48, 1660, 1976.
16. Elfers, L. A. and Decker, C. E., *Anal. Chem.,* 40, 1658, 1968.
17. Hrabeczy-Pall, A., Toth, K., Pungor, E., and Vallo, F., *Anal. Chim. Acta,* 77, 278, 1975.
18. Toth, K. and Pungor, E., *Anal. Chim. Acta,* 51, 221, 1970.
19. Mazor, L., *Analytical Chemistry of Organic Halogen Compounds,* Pergamon Press, Oxford, 1975.
20. Frant, M. S. and Ross, J. W., *Science,* 154, 1553, 1966.
21. Homolo, A. and James, R. O., *Anal. Chem.,* 48, 776, 1976.
22. Radic, N., *Analyst,* 101, 657, 1976.
23. Pioda, L. A. R., Stankova, V., and Simon, W., *Anal. Lett.,* 2, 655, 1969.
24. Ruzicka, J., Hansen, E. H., and Tjell, J. C., *Anal. Chim. Acta,* 67, 155, 1973.
25. Amman, D., Bissig, R., Cimerman, Z., Fiedler, U., Guggi, M., Morf, W. E., Oehme, M., Osswald, H., Pretsch, E., and Simon, W., in *Ion and Enzyme Electrodes in Biology and Medicine,* Kessler, M., Clark, L. C., Jr., Lubbers, W., Silver, I. A., and Simon, W., Eds., Urban and Schwarzenberg, Munich, 1976, 22.

26. Scholler, R. P. and Simon, W., *Chimia,* 24, 372, 1970.
27. Fritz, I., Nagy, G., Fodor, L., and Pungor, E., *Analyst,* 101, 439, 1976.
28. Hansen, E. H., Lamm, C. G., and Růžička, J., *Anal. Chim. Acta,* 59, 403, 1972.
29. Růžička, J. and Hansen, E. H., *Anal. Chim. Acta,* 63, 115, 1973.
30. Hansen, E. H. and Růžička, J., *Anal. Chim. Acta,* 72, 365, 1974.
31. Frant, M. S. and Ross, J. W., *Anal. Chem.,* 40, 1169, 1968.
32. Liberti, A. and Mascini, M., *Anal. Chem.,* 41, 676, 1969.
33. Peters, M. A. and Ladd, D. M., *Talanta,* 18, 655, 1971.
34. Selig, W., *Mikrochimica Acta,* 87, 1973.
35. Smith, M. J. and Manahan, S. E., *Anal. Chem.,* 45, 836, 1973.
36. Hulanicki, A. and Trojanowicz, M., *Anal. Chim. Acta,* 68, 155, 1974.
37. Boch, R. and Puff, H. J., *Z. Anal. Chem.,* 240, 381, 1968.
38. Orion Research, Inc., Newsletter, 2, 42, 1970.
39. Bailey, P. L. and Pungor, E., *Anal. Chim. Acta,* 64, 423, 1973.
40. Horvai, G., Tóth, K., and Pungor, E., *Anal. Chim. Acta,* 82, 45, 1976.
41. Fuchs, C., Dorn, D., Fuchs, C. A., Henning, H. V., McIntosh, C., and Scheler, F., *Clin. Chim. Acta,* 60, 157, 1975.
42. Gran, G., *Analyst,* 77, 661, 1952.
43. Westcott, C. C., *Anal. Chim. Acta,* 86, 269, 1976.
44. Van der Meer, J. M. and Smit, J. C., *Anal. Chim. Acta,* 83, 367, 1976.
45. Van den Winkel, P., Mertens, J., and Massart, D. L., *Anal. Chem.,* 46, 1765, 1974.
46. Fleet, B. and Ho, A. Y. W., in *Ion-selective Electrodes,* Pungor, E., Ed., Akademiai Kiado, Budapest, 1973, 17.
47. Nagy, G., Feher, Z., Tóth, K., and Pungor, E., *Hung. Sci. Instrum.,* 41, 27, 1977.
48. Fleet, B. and Ho, A. Y. W., *Anal. Chem.,* 46, 9, 1974.
49. Nagy, G., Tóth, K., and Pungor, E., *Anal. Chem.,* 47, 1460, 1975.
50. Nagy, G., Feher, Z. Tóth, K., and Pungor, E., *Anal. Chim. Acta,* 91, 87, 1977.
51. Nagy G., Feher, Z., Tóth, K., and Pungor, E., *Anal. Chim. Acta,* 91, 97, 1977.
52. Pungor, E., Feher, Z., and Nagy, G., *Hung. Sci. Instrum.,* 18, 37, 1970.
53. Nagy, G., Feher, Z., and Pungor, E., *Anal. Chim. Acta,* 52, 74, 1970.
54. Pungor, E. and Tóth K., in *Vom Wasser,* Husmann, W., Ed., Vol 42, Verlag Chemie, Weinheim, Germany, 1974, 43.
55. Pungor, E., Tóth, K., and Nagy, G., *Hung, Sci. Instrum.,* 35, 1, 1975.
56. Pungor, E., Tóth, K., and Nagy, G., in *Ion and Enzyme Electrodes in Biology and Medicine,* Kessler, M., Clark, L. C., Jr., Lubbers, W., Silver, I. A., and Simon, W., Eds., Urban and Schwarzenberg, Munich, 1976, 56.
57. Tóth, K., Nagy, G., Feher, Z., and Pungor, E., *Analyst,* 99, 699, 1974.
58. Tóth, K., Nagy, G., Feher, Z., and Pungor, E., *Z. Anal. Chem.,* 282, 379, 1976.
59. Eswara Dutt, V. V. S. and Mottola, H. A., *Anal. Chem.,* 47, 357, 1975.
60. Růžička, J. and Hansen, E. H., *Anal. Chim. Acta,* 78, 145, 1975.
61. Růžička J. and Stewart, J. W. B., *Anal. Chim. Acta,* 79, 79, 1975.
62. Stewart, J. W. B., Růžička, J., Bergamin Filho, H., and Zagatto, E. A., *Anal. Chim. Acta,* 81, 371, 1976.
63. Růžička, J., Stewart, J. W. B., and Zagatto, E. A., *Anal. Chim. Acta,* 81, 387, 1976.
64. Stewart, J. W. B. and Růžička, J., *Anal. Chim. Acta,* 82, 137, 1976.
65. Hansen, E. H. and Růžička, J., *Anal. Chim. Acta,* 87, 353, 1976.
66. Eswara Dutt, V. V. S., Eshander-Hanna, A., and Mottola, H. A., *Anal. Chem.,* 48, 1207, 1976.
67. Růžička, J., Hansen, E. H., and Zagatto, E. A., *Anal. Chim. Acta,* 88, 1, 1977.
68. Hansen, E. H., Růžička, J., and Rietz, B., *Anal. Chim. Acta,* 89, 241, 1977.
69. Mazor, L., Papay, K. M., and Klatsmanyi, P., *Talanta,* 10, 557, 1963.
70. Schoniger, W., *Mikrochim. Acta,* 123, 1955.
71. Hassan, S. S. M., *Z. Anal. Chem.,* 266, 272, 1973.
72. Selig, W., *Z. Anal. Chem.,* 249, 30, 1970.
73. Potman, W. and Dahmen, E. A. M. F., *Mikrochim. Acta,* 303, 1972.
74. Thomas, J. and Clusroter, H. J., *Anal. Chem.,* 46, 1321, 1974.
75. Baker, R. L., *Anal. Chem.,* 44, 1326, 1972.
76. Bremmer, J. M., *Commun. Soil Sci. Plant Anal.,* 3, 159, 1972.
77. Buckee, H. K., *J. Inst. Brewing,* 80, 291, 1974.
78. Mertens, J., Van den Winkel, P., and Massart, D. L., *Anal. Chem.,* 47, 522, 1975.
79. Morie, G. P., Ledford, C. J., and Glover, C., *Anal. Chim. Acta,* 60, 397, 1972.
80. Guilbault, G. G., Smith, R. K., and Montalvo, J. G., Jr., *Anal. Chem.,* 41, 600, 1969.
81. Katz, S. A. and Rechnitz, G. A., *Z. Anal. Chem.,* 196, 248, 1963.

82. Rogers, D. S. and Pool, K. H., *Anal. Lett.*, 6, 801, 1973.
83. Hansen, E. H. and Růžička, J., *Anal. Chim. Acta*, 72, 353, 1974.
84. Selig, W., *Mikrochim. Acta*, 3, 349, 1973.
85. Christova, R., Ivanova, M., and Novkirishka, M., *Anal. Chim. Acta*, 85, 301, 1976.
86. Nagy, G., Von Storp, L. H., and Guilbault, G. G., *Anal. Chim., Acta*, 66, 443, 1973.
87. Honda, S., Sudo, K., Kakehi, K., and Takiura, K., *Anal. Chim. Acta*, 77, 274, 1975.
88. Hassan, S. S. M., *Z. Anal. Chem.*, 270, 125, 1974.
89. Hadjiioannou, T. P., Koupparis, M. A., and Efstathiou, C. E., *Anal. Chim. Acta*, 88, 281, 1977.
90. Altinata, A. and Pekin, B., *Anal. Lett.*, 6, 667, 1973.
91. Radic, N. J., *Analyst*, 101, 657, 1976.
92. Papastathopoulos, D. S., and Rechnitz, G. A., *Anal. Chem.*, 47, 1792, 1975.
93. Llenado, R. A. and Rechnitz, G. A., *Anal. Chem.*, 46, 1109, 1974.
94. Llenado, R. A. and Rechnitz, G. A., *Anal. Chem.*, 45, 826, 1973.
95. Hjemdahl-Monsen, C. E., Papastathopoulos, D. S., and Rechnitz, G. A., *Anal. Chim. Acta*, 88, 253, 1977.
96. Hussein, W. R. and Guilbault, G. G., *Anal. Chim. Acta*, 76, 183, 1975.
97. Hussein, W. R., Von Storp, H., and Guilbault, G. G., *Anal. Chim. Acta*, 61, 89, 1972.

INDEX

A

B

C

D

E

F

G

H

I

N

O

P

R

S

T

U

V

W